INSTRUCTOR'S MANUAL TO ACCOMPANY

INTRODUCTION TO
FORESTRY
SCIENCE *Third Edition*

Join us on the web at

www.cengage.com/community/agriculture

INSTRUCTOR'S MANUAL TO ACCOMPANY

INTRODUCTION TO
FORESTRY
SCIENCE *Third Edition*

L. DeVere Burton

DELMAR
CENGAGE Learning

Australia • Brazil • Japan • Korea • Mexico • Singapore • Spain • United Kingdom • United States

DELMAR
CENGAGE Learning·

Instructor's Manual to Accompany
Introduction to Forestry Science, Third Edition
L. DeVere Burton

Vice President, Editorial: Dave Garza

Director of Learning Solutions: Matthew Kane

Senior Acquisitions Editor: Sherry Dickinson

Managing Editor: Marah Bellegarde

Senior Product Manager: Christina Gifford

Editorial Assistant: Scott Royael

Vice President, Marketing: Jennifer Baker

Marketing Director: Debbie Yarnell

Marketing Manager: Erin Brennan

Marketing Coordinator: Erin DeAngelo

Senior Production Director: Wendy Troeger

Production Manager: Mark Bernard

Senior Content Project Manager: Elizabeth Hough

Senior Art Director: David Arsenault

For product information and technology assistance, contact us at
Cengage Learning Customer & Sales Support, 1-800-354-9706

For permission to use material from this text or product, submit all requests online at **www.cengage.com/permissions**. Further permissions questions can be e-mailed to **permissionrequest@cengage.com**

Library of Congress Control Number: 2011941346

ISBN-13: 978-1-111-30840-7
ISBN-10: 1-111-30840-3

Delmar
5 Maxwell Drive
Clifton Park, NY 12065-2919
USA

Cengage Learning is a leading provider of customized learning solutions with office locations around the globe, including Singapore, the United Kingdom, Australia, Mexico, Brazil, and Japan. Locate your local office at: **international.cengage.com/region**

Cengage Learning products are represented in Canada by Nelson Education, Ltd.

To learn more about Delmar, visit **www.cengage.com/delmar**
Purchase any of our products at your local college store or at our preferred online store **www.cengagebrain.com**

Notice to the Reader
Publisher does not warrant or guarantee any of the products described herein or perform any independent analysis in connection with any of the product information contained herein. Publisher does not assume, and expressly disclaims, any obligation to obtain and include information other than that provided to it by the manufacturer. The reader is expressly warned to consider and adopt all safety precautions that might be indicated by the activities described herein and to avoid all potential hazards. By following the instructions contained herein, the reader willingly assumes all risks in connection with such instructions. The publisher makes no representations or warranties of any kind, including but not limited to, the warranties of fitness for particular purpose or merchantability, nor are any such representations implied with respect to the material set forth herein, and the publisher takes no responsibility with respect to such material. The publisher shall not be liable for any special, consequential, or exemplary damages resulting, in whole or part, from the readers' use of, or reliance upon, this material.

Printed in Milton Keynes UK
1 2 3 4 5   29 28 27 26 25

TABLE OF CONTENTS

Section One

INTRODUCTION TO FORESTRY SCIENCE: GETTING ACQUAINTED WITH THE FOREST   1

CHAPTER 1   INTRODUCTION TO FORESTRY   2

CHAPTER 2   NORTH AMERICAN FOREST REGIONS   6

Section Two

FOREST SAFETY   10

CHAPTER 3   SAFETY IN THE FOREST   11

Section Three

FOREST MANAGEMENT   15

CHAPTER 4   SILVICULTURE   16

CHAPTER 5   MEASUREMENT OF FOREST RESOURCES   20

CHAPTER 6   HARVEST AND REFORESTATION PRACTICES   25

CHAPTER 7   FIRE AND THE FOREST   30

CHAPTER 8   WILDLIFE AND THE FOREST   34

CHAPTER 9   WATER QUALITY AND WATERSHED MANAGEMENT   38

CHAPTER 10   THE ROLE OF GOVERNMENT IN FORESTRY   42

Section Four
FOREST PRODUCTS   46

CHAPTER 11   WOOD CONSTRUCTION MATERIALS   47

CHAPTER 12   SPECIALTY FOREST PRODUCTS   52

CHAPTER 13   PLANTATION PRODUCTS AND PRACTICES   56

Section Five
NEW DIRECTIONS AND TECHNOLOGIES IN FORESTRY   60

CHAPTER 14   URBAN FORESTRY   61

CHAPTER 15   SPACE-AGE FOREST TECHNOLOGIES   66

Section Six
DENDROLOGY: THE SCIENTIFIC STUDY OF TREES   70

CHAPTER 16   PHYSIOLOGY OF TREES   71

CHAPTER 17   FOREST ECOLOGY   75

CHAPTER 18   DISEASES AND PESTS OF TREES   79

CHAPTER 19   ANATOMY AND CLASSIFICATION OF TREES   84

Section Seven
TREES OF THE FOREST   88

CHAPTER 20   TREE IDENTIFICATION   89

Section One

INTRODUCTION TO FORESTRY SCIENCE:

Getting Acquainted with the Forest

Chapter 1

INTRODUCTION TO FORESTRY

OBJECTIVES

After completing this chapter, you should be able to

- identify important forest products that contribute to the comfort and health of people and the economies of nations

- describe the kinds of plants that compose the vegetative strata found in a forest environment

- list the major life forms that contribute to the biological value of a forest

- suggest some natural functions of a forest that affect its biological value

- describe how a watershed functions and explain why a forested watershed is superior to a watershed that lacks forest plant cover

- identify ways that forest environments contribute to stable populations of wild animals

- distinguish between renewable resources and nonrenewable resources

- account for the major uses of forest resources in the United States

- list ways that forest products such as wood and other biomass materials are used as sources of energy

- explain the multiple-use concept of management for public lands

TEACHER BACKGROUND INFORMATION

Forests stabilize many of the processes that occur on the planet, such as the water cycle, remediation of pollution, and the life cycles of many wild organisms. The world would be very different without forests, and many living creatures would be unable to survive. It is important to care for our forest resources in ways that will preserve their productivity and the environments they create.

Renewable resources are those considered to be replaced within a few hundred years. For example, it takes longer to produce a redwood tree than it does to produce a cottonwood tree, but each is renewed in its time. It is possible for coal and petroleum to be replaced, but it requires thousands and perhaps millions of years for this to happen. For this reason, for all practical purposes, they are nonrenewable resources.

INSTRUCTIONAL STRATEGIES

Bring some products to the classroom such as oil, coal, plywood, charcoal, and so on, and ask the students to identify whether each product is renewable or nonrenewable. Discuss why they are renewable or nonrenewable, and explain generally how renewable and nonrenewable resources affect the lives of people.

LEARNING ACTIVITIES

1. Conduct an inventory of the forest products used in your school classroom and laboratory facility. Then have each student conduct a forest products inventory in his or her home. Create a master list of all forest products that class members identified. Use the list to design a classroom bulletin board focusing on forest products.

2. Visit a park or forested area near your school. Make a chart that lists "renewable resources" and "nonrenewable resources" as main topic headings. List each natural resource observed on the field trip under one of these headings. Call on your students to explain why a particular natural resource should be listed in the category he or she has chosen.

3. Invite a guest speaker from a city, state, or national agency who works with trees and forests. Prepare the students to ask questions about tree species in the local area to include how the trees are used, which species are most important locally, what products are obtained from trees in the region, and what skills are needed to work in the industry.

4. Make a list of the most valuable trees in your area. Take time to do an Internet search for the trees of choice to learn as much as possible about each tree. Each student should contribute one tree to the list. Students should be prepared to discuss why their choices should be on the list. Once the list is complete, narrow it down to the "top 10" trees based on the information the students have gathered.

QUESTIONS FOR DISCUSSION AND REVIEW

Essay Questions

1. List some important products obtained from forests that contribute to the health and comfort of people.

   Forests clean the environment by removing impurities from air and water. They also add oxygen to the atmosphere, and they provide medicines, fuels, solvents, paper products, and many other products that contribute to the health and comfort of people.

2. How do forest products contribute to a nation's economy?

   Thirty percent of the land area of the world is forested, and many valuable products are harvested from forest resources. The value of these products contributes to the economies of many countries. Forest products are also renewable resources that will be available to every generation for as long as forests are managed for sustained harvests.

3. What kinds of plants make up the vegetative strata of a forest environment?

The vegetative strata of a forest include many kinds of plants. The canopy consists of tall trees, and the understory consists of smaller trees. The shrub layer consists of small woody plants and bushes, and the herb layer is made up of ferns, grasses, and flowering plants. Decaying plants are found on the forest floor.

4. Name some of the life forms that contribute to the biological value of a forest.

Among the life forms that contribute to biological value are plants, insects, spiders, mammals, birds, amphibians, reptiles, fish, and microscopic life forms.

5. Identify some naturally occurring forest functions that contribute to the biological value of a forest.

Among the naturally occurring functions of forests that contribute to biological value is each of the living organisms. Forests also have effects on climate, watersheds, water temperature, and soils.

6. Describe how a watershed functions, and explain why a forested watershed is superior to a watershed that lacks forest plant cover.

A watershed is a geographical area that absorbs precipitation into the soil, from which it is released into streams and rivers in a fairly uniform flow. The presence of trees in the watershed slows the snow melt and increases the capacity of soil to absorb water.

7. Explain how a healthy forest environment contributes to stable populations of wild animals.

Forests provide habitat for many kinds of wild animals. They provide food and shelter for them, as well as protection from natural enemies. Each of these functions contributes to the stability of animal populations.

8. Distinguish between renewable resources and nonrenewable resources.

Renewable resources are those that are capable of regrowth following their use. A nonrenewable resource, such as coal or petroleum, cannot be replenished once it is consumed.

9. What are some major uses of forest products in the United States?

Major uses of forest products in the United States include lumber (42 percent), pulpwood (28 percent), fuelwood (20 percent), and plywoods and veneers (8 percent). Other uses account for 2 percent of forest product uses.

10. How is wood and other biomass materials converted to energy sources?

Wood and other biomass materials are converted to energy sources such as wood pellets, coke, or chips that are burned to produce energy in the form of heat. The heat thus produced is often converted to electricity at electrical generating plants.

11. Explain the multiple-use concept of management for lands owned by the public.

The multiple-use concept of forest management provides opportunities to use forest resources for purposes such as logging, grazing, mining, recreation, and wildlife.

MULTIPLE-CHOICE QUESTIONS

1. What percent of the land area in the world is forest land?

   A. 20%   C. 40%

   B. <u>30%</u>   D. 50%

2. Broadleaf trees generally represent which of these strata in a forest?

   A. <u>canopy</u>   C. understory

   B. herb layer   D. shrub layer

3. A measurement that takes into account the living organisms in a forest and natural forest functions is called:

   A. forensic value   C. economic value

   B. geological inventory   D. <u>biological value</u>

4. A watershed is a:

   A. <u>region in which precipitation is readily absorbed into the soil</u>

   B. water treatment facility in which pollutants are removed from water

   C. region where water has great difficulty penetrating into the soil

   D. municipal water storage structure

5. A forest is considered a natural resource that is:

   A. permanent   C. <u>renewable</u>

   B. nonrenewable   D. expendable

6. Which of the following uses consumes the greatest amount of harvested timber?

   A. plywood   C. <u>lumber</u>

   B. paper products   D. fuelwood

7. The production of lumber in the United States during the last century has:

   A. increased slightly   C. decreased

   B. <u>remained constant</u>   D. tripled

8. The sale of timber products accounts for what percentage of the gross national product (GNP) in the United States?

   A. <u>6%</u>   C. 11%

   B. 2%   D. 21%

9. How much of the annual harvest of hardwood trees is used for fuelwood?

   A. one-tenth   C. two-thirds

   B. one-fourth   D. <u>nearly half</u>

10. Electricity that is produced using harvested plant materials as fuel to produce heat is called:

    A. hydropower   C. <u>biomass power</u>

    B. nuclear power   D. induced power

11. A forest management strategy that provides public access to forest lands for such activities as grazing, mining, logging, and recreation is called:

    A. conservation reserve   C. selective resource management
       program

    B. resource isolation doctrine   D. <u>multiple-use</u>

Chapter 2

NORTH AMERICAN FOREST REGIONS

TEACHER BACKGROUND INFORMATION

A good book on tree identification and/or a computer program that includes a key to tree identification will be useful to you in teaching this unit. Work with the key to be sure you can use it properly to identify the trees you are likely to encounter in your region. Keep in mind that new tree hybrids confuse the identification of trees and that tree varieties are found in landscapes and parks that would not naturally be found in the region.

INSTRUCTIONAL STRATEGIES

Discuss the value and importance of diversity in the populations of trees. Consider the many kinds of products that come from trees, including medicines, spices, naval stores, and so on, and point out to the students that these products all come from different species of trees. Consider the many additional species found in tropical rain forests, and discuss the implications to medicine and other human needs if these forests are destroyed before we learn their true value.

LEARNING ACTIVITIES

1. Obtain slides or video materials from a biological supply house or other vendor that students can use to learn to identify common trees. Teach students to use the key in a field guide to identify unknown trees.

2. Either as individual students or as a class, visit an arboretum. Obtain permission to gather samples of leaves that can be pressed and dry-mounted. Other leaf samples might be gathered from the trees located in a local community. Prepare, properly label, and display the leaf collections for a class grade or for extra credit.

3. Classify local trees according to leaf type as discussed in this chapter. Plant materials can be obtained and brought into the class, or the teacher may conduct a walking field trip in the vicinity of the school.

QUESTIONS FOR DISCUSSION AND REVIEW

Essay Questions

1. Name the regional forests of North America and locate them on a map.

   The regional forests of North America include the Northern Coniferous Forest, Northern Hardwoods Forest, Central Broadleaved Forest, Southern Forest, Bottomlands Hardwoods Forest, Rocky Mountain Forest, and Pacific Coast Forest. Note the locations of each of these forests on a map.

2. What are the most important species of trees in each regional forest?

   - Northern Coniferous Forest: spruces, pines, tamarack
   - Northern Hardwoods Forest: beech, maple, hemlock, birch, white pine
   - Central Broadleaved Forest: elm, sweetgum, ash, sycamore, black oaks, black walnut
   - Southern Forest: loblolly pine, slash pine, shortleaf pine, longleaf pine, oak, hickory
   - Bottomlands Hardwoods Forest: sweetgum, oaks, pecan, eastern cottonwood, green ash, baldcypress, Atlantic white cedar, pond pine
   - Rocky Mountain Forest: pines, Douglas fir, larch, western red cedar
   - Pacific Coast Forest: ponderosa pine, redwood, pines, Douglas fir

3. Explain the concept of biological succession and distinguish between primary succession and secondary succession.

   Biological or ecological succession is the process by which the plant life evolves from pioneer species to climax species as the environment changes to support more dominant plants. For example, the progression might evolve from weeds to shrubs and bushes, inferior trees, and softwood or hardwood trees. Primary succession occurs when plants become established in an environment for the first time. Secondary succession occurs when an environment supports only those organisms that were found naturally in an earlier stage of succession.

4. What are the characteristics that distinguish conifer, deciduous, and evergreen trees from one another?

The distinguishing characteristics among conifers, deciduous, and evergreen trees are as follows:

- Deciduous trees shed their leaves during part of each year.
- Conifers are trees that produce seeds in cones.
- Evergreens are trees that retain their leaves throughout the year.

5. Identify some characteristics of the Northern Coniferous Forest that may account for its low level of production of forest products in comparison with other North American forests.

Characteristics of the Northern Coniferous Forest that may account for low production levels include short growing seasons, relatively poor and infertile soils, and poor soil drainage.

6. List some important products besides wood and paper that are obtained from forests.

Important forest products besides wood and paper include fuels, maple syrup, chemicals, medicines, naval stores, and fruits and nuts.

7. Why is it important to harvest trees in a timely manner once they have reached maturity?

Trees should be harvested in a timely manner once they reach maturity because heartwood often begins to decay in overmature trees, causing a high incidence of weakness and death among the trees.

8. Define the term *silviculture* and list some examples of important silviculture practices.

Silviculture is the management of forests and their environments to establish, cultivate, and promote the growth and harvesting of trees for commercial purposes. Silviculture practices include planting; controlling weeds, insects, and pests; and pruning, thinning, and harvesting of trees.

9. How is the geological formation known as an *alluvial fan* formed?

An alluvial fan is a geological deposit that occurs as water carries silt, sand, and gravel to an area during flood stage, where the materials are deposited as the water flow slows. Huge gravel and soil deposits occur as the cycle is repeated over hundreds or thousands of years.

10. What characteristics of the Pacific Coast Forest account for its high production of wood products in comparison with other forests?

Characteristics of the Pacific Coast Forest that account for higher productivity than other forests include an abundance of moisture and a climate that is favorable to growth.

MULTIPLE-CHOICE QUESTIONS

1. A natural process that occurs as higher-order plants replace lower-order plants in an environment is called:

   A. primary succession   C. biological succession

   B. secondary succession   D. succession sequence

2. The population of plants that occupies an environment when the succession of plant species is complete and the plant population has stabilized is known as a:

   A. climax community   C. pioneer species

   B. boreal forest   D. terminal forest

3. A conifer is a tree that:

   A. is called an evergreen   C. sheds its leaves or needles every year

   B. <u>produces seeds in cones</u>   D. produces acorns

4. Which important commercial hardwood tree was eliminated from commercial production by a parasitic fungus?

   A. Yellow Birch   C. American Beech

   B. Black Spruce   D. <u>American Chestnut</u>

5. The most abundant and valuable trees in the Central Broadleaved Forest are the:

   A. <u>oaks</u>   C. pines

   B. walnuts   D. hickories

6. Which destructive force became a major problem on native forest lands in the southern and eastern regions of the United States soon after the land was cleared for farming?

   A. hurricanes   C. climate changes

   B. <u>soil erosion</u>   D. flooding

7. Management of forests and their environments for the commercial production and harvest of trees for lumber and other wood products is called:

   A. <u>silviculture</u>   C. horticulture

   B. reforestation   D. forest regeneration

8. Which tree found in the Bottomland Hardwoods Forest is *not* a conifer?

   A. <u>Sweetgum</u>   C. Baldcypress

   B. Atlantic White-Cedar   D. Pond Pine

9. Which forest type consists of pioneer species of trees?

   A. Cypress-Tupelo   C. <u>Cottonwood/Willow</u>

   B. Oak/Hardwoods   D. Hickory-Walnut

10. Cypress and Tupelo trees are adapted to environments that are covered with water during much of the year. Which term best describes these trees?

   A. <u>aquatic</u>   C. pioneer

   B. evergreen   D. pulpwood

Section Two

FOREST SAFETY

Chapter 3

SAFETY IN THE FOREST

OBJECTIVES

After completing this chapter, you should be able to

- describe the type of clothing and personal gear that forest workers should use to provide for their safety

- distinguish between noise intensity and noise duration

- define the term *decibel* and explain how this measurement is related to the safety of workers in the forest and wood products industries

- discuss ways that high-quality tools and equipment contribute to the safety of forest industry workers

- explain how training contributes to the safety of workers in the forest industry

- list some examples of safe work habits that are known to protect forest workers

- discuss the importance of workers' demonstrating respect for the power of the machines and equipment they operate as they do their work

- describe some safety practices that contribute to the safe operation of a chain saw

- list some work habits that contribute to worker safety as timber is harvested and transported

- identify some safety practices that have proven to protect workers in timber processing mills

TEACHER BACKGROUND INFORMATION

This unit will be easier to teach if you have experience in the forest and wood products industry. If this is not the case, it will be helpful to view some industry video materials or visit a wood products processing facility before you teach the unit.

INSTRUCTIONAL STRATEGIES

Make arrangements to bring a full set of safety clothing and equipment to the classroom on the day this unit is introduced. Have a student put them on and remain out of sight until the lesson begins. As the lesson begins, the student should saunter into the room with a sign around his or her neck that says: "I'm a logger." Discuss each item of clothing and equipment with the class.

LEARNING ACTIVITIES

1. Obtain a safety training video for forestry from the National Safety Council, a state industrial commission, state library, or another source in the public domain. Show the video to the students and have them identify each unsafe situation or condition that might result in an accident. Discuss ways that the accidents might have been prevented.

2. Invite a guest speaker to come to the school to speak on the topic of "Worker Safety in Forestry Careers." The speaker should be someone who has had experience in the industry, such as a logging contractor, aircraft pilot, firefighter, or mill worker. Suggestions for speakers could also include an extension forester, a representative of a wood product industry, a supervisor from a wood processing mill, or someone from another forestry career. Invite the speaker to spend most of his or her time on positive suggestions for creating safe working conditions, with just enough good examples of consequences (for unsafe practices) to hold the interest of the students.

QUESTIONS FOR DISCUSSION AND REVIEW

Essay Questions

1. What kinds of clothing and personal protective gear should those who work in outdoor forestry careers wear?

   Clothing should be worn that is capable of providing some protection to forest workers. Proper clothing should include a hard hat, safety glasses, hearing protection, long-sleeved work shirt, leather gloves, durable work pants, and steel-toed shoes.

2. What is the relationship between noise intensity and noise duration as they affect hearing loss?

   Noise intensity is the amount of energy that is in the sound waves of a noise source. Noise duration is how long the noise lasts. A combination of high noise intensity (decibels) and sustained noise duration causes hearing damage.

3. What is the unit of measurement for noise intensity? How is it measured?

   The unit of measurement for noise intensity is the decibel, or dB. Decibels are measured with a noise or decibel meter.

4. How does the quality of tools and equipment contribute to the safety of forest industry workers?

   High-quality tools are important to forest industry workers because these tools contribute to worker safety by functioning properly. A dull cutting tool requires extra effort to complete a task. As the extra effort becomes uncomfortable to the worker, the ability to control the tool is diminished. Loss of control over tools and equipment contributes to accidents and injuries to the workers who use them.

5. Discuss the importance of safety training as it relates to workers in forestry careers.

Training is probably the most important activity contributing to forest safety. This is because safe work habits are learned. Most people develop safety habits by learning from the mistakes of others. Safety training is an important aspect of training for a career in the forest industry.

6. What is meant by "safe work habits"? List some examples.

A safe work habit is a way of responding in performing a task. It begins by thinking through each work situation in an effort to identify potential hazards. This is followed by selecting ways to approach dangerous work situations that will reduce the risk of injury. Examples include cutting away from the body when cutting tools are used and keeping one's body out of a direct line with moving parts.

7. Why is it important for workers to have respect for the power of the machines and equipment that they use?

Workers must learn to recognize the power of the machines and equipment used in the forest industry because mechanical devices apply power to a task from the time the power is turned on until it is turned off. A machine cannot tell whether it is cutting a piece of wood or a human hand—it just keeps on working. The only prevention for this kind of accident is for the people who operate the machines to remain in total control of the machines and to avoid risky situations.

8. What does a worker need to know in order to operate a chain saw safely?

The most dangerous part of a chain saw is the moving chain on which the blades are mounted. It is important to keep the blade of the saw pointed away from the feet and legs while the saw is operated. The throttle should never be locked in the "on" position except during starting. The saw teeth should be kept sharp, and the chain should be allowed to slow down between cuts. Every precaution should be taken to avoid contact with the moving teeth of a chain saw, and care must be taken to avoid being burned by the exhaust.

9. List some work habits that contribute to the safety of workers as they harvest and transport logs.

Loggers must be aware of the direction the tree is likely to fall as they cut down each tree. This will allow them to get in the position of greatest safety as the tree crashes to the ground. They should avoid a position that is in a direct line with the trunk of the tree by stepping back and to one side. This is because a tree sometimes bounces directly backward toward the stump. Wind direction should be taken into account in predicting where the tree will fall. Care must be taken in removing limbs to make sure the log does not roll on the worker.

Workers who transport logs must use care in handling them because their weight can easily crush a worker. Drivers must be careful to avoid tipping their loads, and brakes should be checked often to avoid brake failure. Care must also be exercised as the logs are unloaded to avoid being injured if they should roll or shift as they are moved and placed in stacks.

10. What are some safety practices that should be used to protect workers in timber processing mills?

Mill workers are in a highly mechanized and hazardous environment where their safety depends on developing safe work habits. Proper clothing is important, but so is a healthy respect for machinery. A mill worker must use caution to avoid being crushed by logs or heavy wood products as they move swiftly along the chains and rollers that are used to transport them. Workers must avoid becoming entangled in the chains, rollers, and other mechanical devices that are found in wood processing mills. Machinery should never be worked on until the power is turned off.

MULTIPLE-CHOICE QUESTIONS

1. Which item would *not* be considered protective clothing for a worker in an outdoor forest environment?

A. wool hat   C. steel-reinforced boots

B. long-sleeved work shirt   D. heavy-duty work pants

2. Which item of personal protective equipment should be worn to protect against a safety hazard that is measured in decibels?

   A. hard hat   C. <u>ear protection</u>
   B. eye protection   D. steel-reinforced boots

3. Which factor makes a positive contribution to worker safety?

   A. indifference to danger   C. worn or damaged tools
   B. too little time to do a job D. <u>high-quality tools</u>

4. Which approach to safety training is most likely to help workers learn from the mistakes of others?

   A. reading the owner's manual C. attending a safety attitude class
   B. watching a training video   D. <u>participation in a mentor training program</u>

5. Which procedure is an example of a safe work habit?

   A. <u>turning off the power before making repairs to a machine</u>
   B. trying to get along with a machine that is not working properly to avoid closing the production line
   C. consulting the design engineer before buying new machinery
   D. placing the highest priority on finding ways to do the job faster

6. Which situation would *not* be considered a safe log-handling practice?

   A. equipping log-handling machines with safety cabs and rolls bars
   B. <u>extending the lift to its full height while carrying logs to new locations</u>
   C. extending the lift just enough to lift the log off the ground as it is moved
   D. using helicopters to transports logs from areas that are sensitive to soil damage

7. Chain saw safety is described in each of the following statements with the *exception* of:

   A. Make sure the saw is equipped with a spark arrester.
   B. Keep the blade of the saw pointed away from the legs and feet of the operator.
   C. <u>Lock the throttle in the "on" position except during starting.</u>
   D. Do not allow the blade of the saw to come in contact with rocks or soil.

8. A logger who is harvesting a tree should move to a position of maximum safety as the tree begins to fall. Where is the safest working position in relation to the stump?

   A. <u>back from the stump and to one side</u>   C. directly behind the stump
   B. to the immediate left of the stump   D. at least 20 feet away from the stump

9. Which condition is a major contributor to accidents related to transporting forest products?

   A. driving too fast for road conditions   C. faulty brakes
   B. top-heavy loads   D. <u>all of the above</u>

10. Which class of injuries is most likely to occur in a wood processing facility?

    A. <u>cuts and bruises</u>   C. head injuries
    B. chemical poisoning   D. burns

Section Three
FOREST MANAGEMENT

Chapter 4

SILVICULTURE

After completing this chapter, you should be able to

- define *silviculture*

- list some important silviculture management practices

- distinguish between natural and artificial methods of regenerating forests

- discuss the advantages of direct seeding or planting seedlings in comparison with natural methods for regenerating forests

- explain the most common methods of producing seedlings for forest regeneration

- describe the steps that should be followed in transplanting a tree seedling

- describe the characteristics of the different growth stages of trees, such as seedling, sapling, pole, and mature tree

- explain why it is necessary to control populations of rodents, especially during the seedling and sapling stages of tree development

- describe some intermediate treatments that are applied to forests

- describe some silviculture practices that are used to improve the growth and quality of trees

- explain how the final use of a tree affects the harvesting method used

TEACHER BACKGROUND INFORMATION

Silviculture is a type of farming in which the crop is seedlings, nursery stock, or wood. As with other crops, there are known cultural practices that have been demonstrated to increase the yield or shorten the growth cycle of the crop. These proven methods are sometimes referred to as "best practices." You can become familiar with these practices through literature published by forestry councils, university extension bulletins, and other industry publications.

INSTRUCTIONAL STRATEGIES

A field trip to a tree farm would be an ideal way to introduce this unit. Tree farms exist in most areas, and it does not matter what kind of a tree farm it is. In rural areas, it is usually possible to locate a forest nursery, Christmas tree plantation, or even a tree farm that is producing mature trees. In urban areas, it will usually be possible to locate a farm used to produce landscaping trees and shrubs. Any of these types of tree farms will work well as a site for a field trip. The farm manager should be asked what, how, and why with regard to silviculture practices observed.

LEARNING ACTIVITIES

1. Obtain seeds and fresh cuttings for a commercial tree species and generate seedlings. Compare the growth of seedlings obtained from vegetative cuttings with the growth of seedlings generated from seeds. Assign students to keep a log book in which they record their observations and their work throughout the project. Assemble the combined data of the entire class and assess the two methods for generating seedlings.

2. Take a field trip to a tree farm and assign the students to prepare a written report that contains at least the following information: (a) the final product or products of the farm, (b) specific cultural practices, (c) handling procedures for each product, (d) product markets and marketing plans, and (e) future plans.

3. Identify one or more varieties of trees that are raised commercially in your area. Evaluate each of them for the following characteristics: (a) Is the variety regenerated by nature or through artificial methods? (b) What is the ideal number of trees to establish per acre or hectare? (c) What is the expected yield at harvest time? (d) What is the estimated value of the trees at harvest time? (e) How soon is a crop of trees of this variety expected to mature?

QUESTIONS FOR DISCUSSION AND REVIEW

Essay Questions

1. What is silviculture?

   Silviculture is the art and science of tree production or tree farming. It involves manipulating the environment to make it favorable to the growth of trees.

2. What are some important silviculture practices, and how do they improve forest production?

   Examples of important silviculture practices include soil testing, seedbed preparation, weed control, disease prevention and control, insect control, thinning, and so forth.

3. How is artificial regeneration of a forest different from natural regeneration?

   Artificial regeneration of a forest occurs when seeds or seedlings are planted at the site of a tree harvest. Natural regeneration of a forest occurs on a site when young trees begin to grow there under natural conditions and no seeds or seedlings are planted.

4. What advantages are gained from direct seeding or planting seedlings in comparison with natural methods of reforestation?

   Advantages of artificial regeneration over natural reforestation methods include uniform stands of trees that are evenly dispersed throughout the area. The forest manager also controls the kinds or species of trees that make up the new forest. Trees that are resistant to diseases and insects can be introduced. Planting in rows makes it possible to use mechanical weed-control methods. Even-aged forests are easier to manage for growth than mixed-age forests.

5. Describe two common methods used to produce tree seedlings, and list the advantages and limitations of each method.

Methods of production used in raising seedlings include:

- Direct seeding: Seeds are planted in rows at a tree nursery, where they are later harvested as bare-root stock in preparation for planting in harvested areas of the forest.
- Containerized seedlings: Seeds are planted in multiplant containers at the rate of two seeds per section. Later, they are thinned to one plant per section. The seedling and the ball of soil surrounding the root mass are planted at the harvest sites.

6. How should the site be prepared for transplanting tree seedlings, and what transplanting procedures should be followed?

The forest site should be prepared to receive tree seedlings by gathering and burning or by mechanically cutting up the forest debris on the soil surface. Hardwood shrubs need to be eliminated prior to planting conifers. The seedbed should be prepared mechanically when possible. At the very least, a small site should be prepared in the area where the seedling will be planted. The hole in which the seedling is transplanted should be large enough to allow the roots to be spread during the planting procedure. Moist soil should be packed firmly around the roots to allow the tree to absorb water and nutrients. The seedling should be planted in a vertical position.

7. Name the stages of growth that trees go through from planting to harvest.

The stages of growth that trees go through from planting to harvest include the seedling, sapling, pole, and mature stages.

8. What are some intermediate treatments that are applied to forests, and what advantages do they contribute to forest health and production?

Intermediate treatments applied to forests include release cuttings (frees saplings from competition from competing species), improvement cuts or liberation (removal of undesirable trees), intermediate cutting or thinning (reduction of stand density), fertilization, salvage cutting (removal of trees damaged by destructive agents such as fire, insects, or diseases), and sanitation cutting (an emergency cutting to stop the spread of diseases and insect populations by removing infected trees).

9. What are some silviculture practices that are used to improve the growth and quality of trees?

Silviculture practices used to improve the growth and quality of trees include pruning, irrigating, applying fertilizer, and thinning.

10. How does the final use of a tree affect the method used to harvest it?

The final use of a tree influences the harvest methods used. For example, trees to be used for pulp by the paper industry are often sheared off at the base with large mechanical jaws. The shearing action crushes the wood near the cut, reducing its value as a lumber product, but it does not interfere with paper quality. The final use of wood can also determine when a tree is harvested. The size of the tree is important in lumber production, but nearly any size of tree can be harvested for pulping purposes.

MULTIPLE-CHOICE QUESTIONS

1. The art and science of tree production is known as:

   A. silvics   C. silviculture

   B. forest regeneration   D. forestry

2. Reproduction of trees from the leaves, stems, or root tissues is called:

   A. <u>vegetative reproduction</u>   C. silviculture

   B. sexual reproduction   D. suckering

3. The natural growth of a young forest following the harvest of mature trees is called:

   A. spontaneous combustion   C. artificial regeneration

   B. arboriculture   D. <u>natural regeneration</u>

4. The process by which seeds begin to sprout and grow is known as:

   A. <u>germination</u>   C. regeneration

   B. gymnosperm   D. angiosperm

5. Young trees that have been removed from the soil in preparation for shipping are known as:

   A. seedlings   C. <u>bare-root stock</u>

   B. saplings   D. poles

6. Some kinds of plants release chemicals into the area close around them that provide small doses of poison to young plants that invade their territory. This defensive plant response is known as:

   A. <u>allelopathic effect</u>   C. herbicide

   B. germicide   D. chemical warfare

7. A harvest method in which the only trees that are harvested are the mature trees and diseased or damaged trees is:

   A. clear-cutting   C. salvage logging

   B. <u>selection cutting</u>   D. girdling

8. A young tree with a diameter of 4–10 inches is a:

   A. sapling   C. <u>pole</u>

   B. seedling   D. log

9. Removal of undesirable older trees from a stand to make sunlight available to young trees is known as a:

   A. <u>liberation</u>   C. stand improvement

   B. cleaning operation   D. sanitation cutting

10. Pruning of trees is a cultural practice that is performed for the purpose of:

   A. harvesting firewood   C. attracting wild animals

   B. <u>improving lumber quality</u>   D. harvesting damaged timber

11. A mechanical timber harvesting method that is used to harvest small- to moderate-sized trees is known as:

   A. intermediate harvest   C. cleaning operation

   B. aerial harvesting   D. <u>whole-tree harvesting</u>

Chapter 5

MEASUREMENT OF FOREST RESOURCES

OBJECTIVES

After completing this chapter, you should be able to

- identify the types of information needed to develop a long-term forest management plan based on sustained yields

- describe the features of the two types of land surveys used in the United States

- explain the relationship of baselines and principal meridians to the initial point location from which each rectangular survey begins

- explain how instruments such as the stereoscope and the planimeter are used in preparing a forest-type map

- name some tools used to estimate the diameter and height of a standing tree, and explain how each tool is used

- define the role of a timber cruiser

- contrast the differences between a 100 percent cruise and a partial cruise, and explain when it is appropriate to use each

- list some assumptions that apply to partial cruises, and explain how these may influence the accuracy of the results

- explain the formula for measuring forest growth, and describe each of the formula's components

- describe the most commonly used methods for scaling logs

TEACHER BACKGROUND INFORMATION

Measurement is an important aspect of forestry, and accuracy in measurement must be achieved if sustained yields of forest products are to be achieved. Students need to learn the value of accurate measurement in order to appreciate the importance of this chapter. Forest surveys require measuring skills in establishing the boundaries of a survey plot. Teachers should familiarize themselves with the correct use of surveying instruments before trying to teach this unit.

INSTRUCTIONAL STRATEGIES

Assign teams of students to stake sample plots as though they were preparing to do a timber cruise. They will require some instruction on the correct use of a surveying instrument in reading degree settings and taking measurements to establish plot boundaries. Each group should have a different size and shape of plot, and each group should check the accuracy of the work of each other group. This exercise may be used to introduce this chapter.

LEARNING ACTIVITIES

1. Cut some cross sections that are 2 to 3 inches thick from several cut trees of different sizes. Assign students to work together in small groups with each group receiving one of the cross sections of a tree. Assume that the tree was harvested last year. Determine the age of the tree at the time it was harvested. Determine some years when drought or other stressful factors affected the tree, and determine some years when growth conditions were favorable. Use pins to mark the annual ring that was deposited when significant events occurred, such as the end of World War II or the year when men walked on the moon.

2. Obtain some aerial photographs or maps of agricultural fields or timber sales areas from a government agency such as the Farm Service Agency or Forest Service. Also obtain some dot-grid sheets from the same agency or purchase some from a forestry supply catalog. Invite a professional forester or soil conservation officer to instruct the class on the methods and purposes for measuring land area. Make sure that each student successfully calculates the land area on his or her field photograph or map.

QUESTIONS FOR DISCUSSION AND REVIEW

Essay Questions

1. Explain the management concept of sustained yield, and identify the kinds of information that are needed to develop a sustained-yield forest management plan.

   The management concept of sustained yield is to determine the production of the forest and harvest the amount of timber each year that the forest can maintain through growth without a net reduction in the inventory of trees. To do this, you need to know the annual growth or yield of the forest and the amount of losses as a result of fire, insects, diseases, and natural disasters.

2. What are the two types of land surveys that are used in the United States? What are the key differences between them?

   The two types of land surveys that have been used in the United States are:

   • Metes and bounds: starting points for surveys were streams, roads, rocks, and other natural features in the landscape. Some of these features have not proved to be permanent.

   • Rectangular survey: an initial point is used for each survey region consisting of a prominent and permanent landscape feature, such as a hill.

3. How is the rectangular survey system organized to identify the locations of property lines and boundaries?

   The rectangular survey is organized as a system of parallel baselines that run east-west and principal meridians that run north-south at 24-mile intervals on each side. Sixteen townships measuring 6 miles on each side are fitted in each set of intersecting parallels and meridians. A range is the east-west location of a township from a principal meridian. Each township is divided into 36 sections, each approximately 1 mile square. Sections are further divided into quarter sections of 160 acres. Each quarter section is divided into four parcels of land, with each parcel called a forty (40 acres).

4. Describe how instruments such as the stereoscope and planimeter are used to create a forest type map.

Forest maps are created with such instruments as the stereoscope and the planimeter. A stereoscope is used to view two different aerial photographs at the same time. When taken from different angles, the view in the stereoscope is a three-dimensional view. This allows technicians to determine the boundaries of different forest types on maps. A planimeter is used to trace a perimeter boundary on a forest map that is drawn to scale to determine land area.

5. How does a timber cruiser determine the height and diameters of live trees?

A timber cruiser determines height and diameters of live trees by converting a dbh measurement to an estimate of diameter or timber volume. Instruments that might be used include a tree caliper, steel tape, Biltmore stick, or similar instrument. Tree height is measured by using an instrument called a hypsometer through which a sighting is taken on the top of the tree at a distance of one chain (66 feet) from the tree. Height can then be calculated using geometric or trigonometric formulas for similar triangles.

6. How does a forest worker determine the age of a live tree?

The age of a live tree can be determined by removing from the tree a thin core of wood that extends to the center of the trunk. An instrument known as an increment borer is used. Age is determined by counting the annual rings of the wood core after it is removed from the tree.

7. What is the difference between a 100 percent timber cruise and a partial timber cruise? When is it appropriate to use each method?

A 100 percent timber cruise is a measurement of every tree in the survey area. A partial cruise is a measurement of a representative sample of the trees to obtain an estimate of the timber volume for all of the trees in the forest. Methods of obtaining the sample include systematic sampling, line plot cruising, and others. When care is used to identify a representative sample, these methods are reasonably accurate.

8. Suggest some sampling and measuring practices that can influence the accuracy of a partial timber cruise.

Sampling and measuring practices that improve the accuracy of a partial cruise include the following:

• Make sure the sample is representative of the entire forest.
• Take enough samples to reduce the chance that errors will occur.
• Make sure the plot size is large enough to tally 15 to 20 trees per plot.

9. Explain the formula for determining forest growth, and describe each of the formula's components.

The formula for determining forest growth is:

$$\text{ingrowth} + \text{survivor growth} - \text{mortality} - \text{cut} = \text{forest growth}$$

• Ingrowth = growth in new trees that were not present in the original survey and trees that were too small to be tallied
• Survivor growth = the difference in timber volume of the trees that survived since the original survey
• Mortality = all the trees that died during the time interval between surveys
• Cut = the measurement taken from the stumps of harvested trees

10. List the steps and explain the procedures that are used to scale logs.

Several different systems are used to scale logs. The steps vary, but they are similar for most scaling methods.

- Measure the diameter of the log on the small end inside the bark.
- Measure the length of the log.
- Determine the number of board feet in a log by consulting a table constructed for this purpose.

Note: Some calculations may require measuring both ends of the log.

MULTIPLE-CHOICE QUESTIONS

1. The type of land survey system used in most areas of the United States, except in some areas of the Southeast, is:

   A. metes and bounds   C. Scribner's rule
   B. <u>rectangular survey</u>   D. Gannet survey

2. What is the name of the survey line that runs east-west?

   A. <u>principal meridian</u>   C. guide meridian
   B. section   D. baseline

3. Each section of land has a surface area of approximately how many acres?

   A. <u>640</u>   C. 160
   B. 320   D. 40

4. A tract of land approximately 6 miles square is called a:

   A. section   C. quarter section
   B. <u>township</u>   D. range

5. A procedure called the dot-grid method uses a clear plastic sheet with evenly spaced dots to:

   A. <u>measure land area</u>   C. make forest type maps
   B. measure tree heights   D. calculate forest basal area

6. A method that uses aerial photographs to make field maps to scale is:

   A. <u>photogrammetry</u>   C. planimetery
   B. altimetric scaling   D. fotometry

7. An instrument that is used to measure the circumference of a tree is called a:

   A. Biltmore stick   C. hypsometer
   B. <u>steel tape</u>   D. timber cruiser

8. The age of a living tree is determined using an instrument called a:

   A. <u>increment borer</u>   C. Biltmore stick
   B. planimeter   D. stereoscope

9. The activity of gathering data for the purpose of estimating the volume of standing timber is called:

   A. slumming   C. <u>cruising</u>
   B. scaling   D. Scribner's rule

10. The timber volume table used by the U.S. Forest Service to measure timber sales is:

    A. Doyle's rule   C. Scribner's rule
    B. <u>Scribner's decimal C</u>   D. International log rule

11. The activity of measuring piled logs to determine their yield in board feet is called:

    A. slumming   C. cruising

    B. <u>scaling</u>   D. scribbing

12. The standard of measurement that is used to express biomass production is:

    A. cubic feet   C. cubic meters

    B. board feet   D. <u>net weight</u>

Measurement Exercise

1. Calculate the number of cords of wood in a stack measuring $4' \times 4' \times 12'$.

   $(4 \times 4 \times 12) \div 128 =$ **1.5 cords**

2. Calculate the number of cords of wood in a stack measuring $6' \times 2' \times 16'$.

   $(6 \times 2 \times 16) \div 128 =$ **1.5 cords**

3. Calculate the number of board feet in a timber measuring $16'' \times 4'' \times 14''$.

   $(16 \times 4 \times 14) \div 12 =$ **74.67 board feet**

4. Calculate the number of board feet in three boards measuring $8'' \times 1'' \times 96''$:

   $(3 \times 8 \times 1 \times 96) \div 144 =$ **16 board feet**

5. Calculate the board foot volume of a log using the International log rule ($V = 0.905 \times ([0.22 \times D \times D] - [0.71 \times D])$) where D = diameter of a section of log 4' long. ***Note:*** This formula includes a factor (0.905) that adjusts for a saw kerf of ¼" per cut.

   A. Log diameter = 7"

      $V = .905 \times ([.22 \times 7 \times 7] - [.71 \times 7] = 9.76 - 4.97 =$ **4.79**

   A. Log diameter = 6.5"

      $V = .905 \times ([.22 \times 6.5 \times 6.5] - [.71 \times 6.5] = 8.41 - 4.61 =$ **3.80**

   B. Log diameter = 6"

      $V = .905 \times ([.22 \times 6 \times 6] - [.71 \times 6] = 7.17 - 4.26 =$ **2.91**

   C. How many total board feet are in the 12' log?

      $4.79 + 3.80 + 2.91 =$ **11.50 board feet**

Chapter 6

HARVEST AND REFORESTATION PRACTICES

After completing this chapter, you should be able to

- identify factors that influence decisions affecting timber harvests
- describe some important components of a timber harvest plan
- distinguish between harvest methods leading to even-aged and uneven-aged forests
- explain how the planned method of forest regeneration affects the selection of a harvest method
- relate the volume of timber harvests to forest growth as it affects sustained yield

- speculate on the reasons the National Environmental Policy Act included a requirement for an environmental impact statement to be filed as part of each timber harvest plan
- list and explain each step that is involved in harvesting timber
- explain the historical relationship between road construction in forests and surface water quality
- evaluate the practice of salvage logging
- describe some methods used to minimize litigation related to timber harvests

TEACHER BACKGROUND INFORMATION

Timber harvests changed in many ways during the last half of the twentieth century. The days are gone when a forest ranger could take an order for a few thousand board feet of timber for a new barn. Fifty years ago, the trees were often marked for harvest within a few days or weeks after the order was placed. Logging permits were issued in a timely manner, and the logs were soon headed for the mill.

In the relatively short span of 50 years, many different rules have been promulgated in the forest industry. Some of these rules are attempts to deal with abuses associated with timber harvests. Other rules have simply evolved over the years with little basis for their existence. Whatever their purposes may be, rules must be addressed in timber harvest plans, and plans must be approved at various levels before a single tree is harvested. In many instances, the development of a harvest plan from start to finish may take several years to complete. Even then, harvests are often challenged and delayed by court actions filed by citizens' organizations with a variety of agendas. To be successfully accomplished, a harvest plan must be able to stand up to criticism and litigation.

INSTRUCTIONAL STRATEGIES

Prepare a demonstration on safe use of a chain saw with emphasis on maintaining this important forest tool in good working order. Students need to understand why oil is mixed with the gasoline and how the blade is kept sharp and serviceable. There is nothing quite like the roar of a chain saw to rivet the attention of a class on this topic, especially when the roar is heard right after the instructor has assigned class members to pretend to be various kinds of trees. Of course, no trees are harvested in this demonstration, but it is an attention getter. Keep in mind that a chain saw should not be operated in an enclosed area because the gasoline engine produces poisonous carbon monoxide fumes.

LEARNING ACTIVITIES

1. Invite a forest owner or representative from the Forest Service to make an illustrated presentation and lead a class discussion on forest harvesting practices. He or she might be asked to bring some examples of old saws and harvesting equipment, along with some modern harvesting tools and equipment. Discuss the following points and any others that might be important in your region of North America:

   • changes in timber harvest practices

   • changes in forest species in a particular forest

   • harvest issues that often lead to litigation

   • laws and regulations and their effects on timber harvesting

2. Make two identical models of a watershed using real soil in a shallow tray. Design it in such a way that it can be tilted to different slopes. Design a water distribution system to be placed at the top of the model and a water collection system to be placed at the lower end of the model. Place a mulch of leaves and plant material on the soil surface, and apply an adequate amount of water through the distribution system to cause water to flow into the collection system. The amount of water used and the rate of flow must always be the same with both experimental models. Collect all of the run-off water and strain it through a filter to collect any silt it may contain. Dry the filter and compare its weight with that of a dry, clean filter. Repeat the experiment using the second model with the only difference being that no plant cover is used on the second model. Compare the results. Repeat the experiment, changing only the slope of the model watershed. Compare results for plant cover versus no plant cover and for differences in slope. Exhibit the model and report your results in a science fair.

QUESTIONS FOR DISCUSSION AND REVIEW

Essay Questions

1. Name some factors that influence decisions affecting timber harvests.

   Factors that influence decisions affecting timber harvests include legislation, regulations, environmental impact statements, occupational safety regulations, and forest management rules.

2. What are the key components in a timber harvest plan?

   The key components of a timber harvest plan include:

   - The method that will be used to regenerate the forest
   - The harvest method that is most compatible with the method of forest regeneration
   - An environmental impact study or statement
   - The amount of timber to be harvested
   - Responses to local concerns

3. List some timber harvest methods that lead to the regeneration of even-aged forests and uneven-aged forests.

   Timber harvest methods that lead to the regeneration of even-aged forests include clear-cutting, the seed-tree method, and the shelterwood method. Selection cutting results in an uneven-aged forest.

4. How is the seed-tree harvest method different from the shelterwood harvest method?

   The seed-tree harvest method leaves mature trees in scattered locations throughout the forest to produce seeds for the next generation of trees. Once the seedlings become established, the seed trees can be harvested. The shelterwood method leaves enough trees in the forest to provide partial shade for the young seedlings after some of the seed trees are harvested. The remaining trees are harvested in a second cutting after the seedlings have become well established.

5. How is the selection-cutting harvest method applied in a forest?

   The selection-cutting harvest method is applied by marking trees for harvest that are near the end of their productive lives. These harvests occur at regular intervals, and care must be taken to harvest only as much timber as the forest can replace by the next harvest.

6. Explain how forest managers determine the amount of timber that should be harvested to establish and maintain a sustained yield.

   In order to establish and maintain a sustained yield of timber, forest managers must determine the amount of timber in the forest inventory and the amount of growth that occurs each year. A harvest cannot exceed the average annual gain in the forest inventory.

7. Why do you think Congress passed the National Environmental Policy Act requiring environmental impact statements to be filed as part of timber harvest plans?

   Congress passed the National Environmental Policy Act requiring environmental impact statements as part of timber harvest plans to ensure that negative impacts of logging activities will be minimized. This is a way to create awareness of the possible effects of habitat destruction on wildlife, potential for soil erosion and pollution of surface water, or loss of native plants, to name only a few.

8. List the main steps involved in harvesting timber, and explain how each step is performed.

   The main steps involved in timber harvesting include the following:

   - Felling the tree: This is done with a chain saw or a mechanical tree feller equipped with a shear or sawhead.

- Limbing: This is done with a mechanical limber or with a chain saw.
- Bucking: A chain saw is used to cut the tree to the proper lengths for hauling or processing.
- Bunching: This operation involves gathering the logs to yards and landings.
- Transporting: Most logs are hauled in trucks to processing plants or railroad shipping yards.

9. What effect does the construction of roads for timber harvests sometimes have on the quality of surface water?

   Road construction in forests must be carefully planned to make sure that the disturbed soil does not erode into the rivers and streams. Silt is the number one pollutant of surface water.

10. What are the positive and negative effects of the forest management practice known as salvage logging?

    Salvage logging allows the remaining value in damaged timber to be recovered when harvests are accomplished within a few months after disease, insect infestation, or damage is incurred. Negative impacts of salvage logging include damage to soil following fires that destroy soil structure.

11. What kinds of data do forest managers include in timber harvest plans to minimize the potential for litigation in the court system?

    Forest managers should include data in harvest plans that support harvests over anticipated litigation issues. In many cases, the key issues that lead to litigation are known long before the harvest plan is developed. Addressing such issues in the harvest plan can speed up the approval of timber sales.

MULTIPLE-CHOICE QUESTIONS

1. The greatest losses of wood in a forest that is past maturity are a result of:

   A. <u>decay</u>   C. fire

   B. flooding   D. insects

2. Which timber harvest method results in an uneven-aged forest?

   A. coppice method   C. <u>selection-cutting method</u>

   B. seed-tree method   D. shelterwood method

3. Which timber harvest method removes all trees from the harvest area?

   A. seed-tree method   C. selection-cutting method

   B. shelterwood method   D. <u>clear-cutting method</u>

4. Which timber harvest method saves enough mature trees in the harvest area to provide shade on the forest floor?

   A. <u>shelterwood method</u>   C. coppice method

   B. seed-tree method   D. selection-cutting method

5. A type of forest regeneration in which the new generation of trees arises from the stumps of the harvested trees is the:

   A. seed-tree method   C. selection-cutting method

   B. <u>coppice method</u>   D. shelterwood method

6. Which government agency has the responsibility to enforce timber harvest regulations requiring environmental impact studies?

   A. NEPA   C. USFS

   B. BLM   D. <u>EPA</u>

7. A problem associated with logging activities such as log skidding and road construction resulting in contamination of surface water is known as:

   A. bucking   C. scaling

   B. <u>siltation</u>   D. skidding

8. A timber harvest practice in which logs are cut into marketable lengths is called:

   A. <u>bucking</u>   C. scaling

   B. siltation   D. skidding

9. A timber harvest operation in which entire trees are processed in the forest to produce biomass is called:

   A. debarking   C. <u>chipping</u>

   B. bucking   D. chaining

10. A type of timber harvest initiated as a result of timber damage caused by insects, fire, or other natural disasters is called:

    A. selection cutting   C. coppice harvesting

    B. <u>salvage logging</u>   D. damage control

Chapter 7

FIRE AND THE FOREST

OBJECTIVES

After completing this chapter, you should be able to

- describe ways that fire is both beneficial and destructive to forests
- identify three key elements that must be present for a fire to occur
- analyze the differences that exist among surface, ground, and crown fires
- list the major causes of destructive forest fires
- explain the effects of wildfires on forests and forest environments
- discuss ways in which prescribed burns may be used to improve the health of forests
- analyze the fire suppression policy of the U.S. Forest Service as it has been implemented in the past and as it exists today
- calculate the rate of spread for a fire at different wind speeds
- describe the indirect attack method of fire suppression
- explain how direct attack methods of fire suppression act on the key ingredients of fire
- evaluate past and present efforts of government and industry to prevent destructive fires in the forests
- assess the effects of a short-duration fire cycle on the health of a forest

TEACHER BACKGROUND INFORMATION

Fire science constitutes an entire discipline of study, and only a few of its topics are introduced in this chapter. A major fire suppression effort is much like a military operation. Large numbers of firefighters are involved, and aircraft are used for surveillance and for dropping water and chemicals on the flames. Smoke jumpers parachute into critical areas to create firebreaks and to extinguish flames. Thousands of meals are served daily, and tons of equipment are hauled to the fire in hundreds of trucks. Vans and buses clog the narrow forest roads as firefighters are transported to and from the fire. This kind of operation requires great organizational skills.

Teachers may want to localize the instruction on fire use and suppression by requesting information from local agencies concerning their fire suppression policies. Many different factors are considered in the writing of a fire suppression policy, and many different policies exist across the country.

INSTRUCTIONAL STRATEGIES

Obtain a training video from an agency engaged in suppressing fires and/or invite a person to be a guest speaker who has been involved as a firefighter. Have the speaker tell some of his or her experiences to the class to illustrate the kind of work that firefighters do, and give the speaker some specific questions ahead of time to which he or she can respond. The teacher should keep the discussion focused by asking key questions that hold the speaker to the subject.

LEARNING ACTIVITIES

1. Obtain a training video that a public agency uses to provide training for fire crews and show it to the class. Invite an experienced firefighter to describe some of his or her experiences with fires to the class. Concentrate on the training that is required before a person is qualified to work as a member of a fire crew.

2. Demonstrate the importance of oxygen to a fire by removing flammable materials and preparing an appropriate demonstration site. Light two candles. After both candles are burning, place an inverted glass jar over one of the candles. Blow gently on the base of the flame on the second candle. Discuss what happened in each case, and ask key questions such as (1) Why was the fire extinguished in the closed container? (2) Why did the fire burn more vigorously when you blew on it?

QUESTIONS FOR DISCUSSION AND REVIEW

Essay Questions

1. In what ways is fire considered to be destructive to forests? How is it beneficial?

   Massive, uncontrolled fires are very destructive to forests because of the intense heat that is generated. Such a fire kills all or most of the trees, destroys the texture of the soil, destroys habitats of endangered species of plants and animals, and brings death to many birds and animals, especially the young. Fire is beneficial when it moves through an area regularly and destroys the debris on the forest floor before it builds up into a major supply of fuel sufficient to support a major fire that kills trees. Fires are beneficial in other ways. They thin young stands of trees, control plants that choke the forest floor, and open up large areas where food plants can grow.

2. What are the three elements that must be present for a fire to occur?

   The three elements that must be present for a fire to occur are the availability and concentration of fuels, heat energy sufficient to raise the fuel to its combustion temperature, and an adequate supply of oxygen to support the fire.

3. How are surface fires, ground fires, and crown fires different from one another?

   The characteristics of surface fires, ground fires, and crown fires are fairly distinct. A surface fire generates only enough heat to burn the layer of twigs, grass, leaves, and dead branches on the forest floor.

A ground fire burns fuels that are on or beneath the ground surface, such as peat or duff. These fires sometimes occur during dry periods, and they are very difficult to extinguish. A crown fire occurs when intense heat is generated, and the fire moves into the tops of the trees, where it burns the foliage. This occurs more often in conifers than in hardwood forests.

4. List the major causes of destructive forest fires.

   The major causes of destructive fires include dry conditions in combination with a buildup of fuel on the forest floor, low humidity, hot seasonal temperatures, wind, and lightning storms. This combination of elements is deadly.

5. Explain how wildfires affect forests and forest environments.

   Wildfires affect forests and forest environments very differently than surface fires, because wildfires cause massive damage and sometimes total destruction to mature trees. Wildlife sustains losses in wildfires, although in the big picture, these losses are usually minimal except for instances where critical habitat for threatened or endangered species is destroyed.

6. What are some ways that prescribed fires are used to improve the health of forests?

   Prescribed fires improve forest health by burning debris from the forest floor before it builds up enough to sustain a fire of high intensity. Prescribed fires thin the forest by reducing the seedling population, thus giving a competitive advantage for water and nutrients to the trees that remain.

7. How effective has the fire suppression policy of the U.S. Forest Service proven to be?

   The fire suppression policy of the U.S. Forest Service has changed in recent years from a no-tolerance policy toward fires to a policy of assessment of the cost of suppressing a fire in comparison with potential losses if it is allowed to burn. This keeps the forest managers more focused on suppressing the fires that threaten the timber and resources of highest value. It is a more effective approach than trying to suppress all fires.

8. Calculate the rate of spread of a fire when the wind is blowing 30 mph in comparison with 10 mph.

   The rate of spread for a fire can be calculated as follows. Given
   $R = W^2$:

   $R = 10^2 = 100$ at wind speed of 10 mph

   $R = 30^2 = 900$ at wind speed of 30 mph

   $900/100 = 9$

   The rate of spread is 9 times as high with 30 mph winds as it is with 10 mph winds.

9. Give some examples of the direct attack and indirect attack methods of fire suppression.

   Examples of direct attack methods of fire suppression include applying fire retardants or water to a fire or using dirt to smother hot embers. An indirect attack uses methods such as establishing firebreaks to stop the forward progress of a fire and setting backfires to deplete the fuel supply.

10. What steps have government agencies and the forest industry taken to prevent destructive fires in the forests?

    The forest industry and government agencies attempt to prevent forest fires by educating the public with radio, television, and printed information on the causes of fires and ways to prevent them. Smokey Bear is an example of a long-term public image encouraging individuals to prevent forest fires by being responsible citizens. Fire danger signs also advise us when fire conditions are high in the forest, and public service announcements on television and radio advise the public of dangerous fire conditions.

11. What effects can be expected on the health of a forest from a short-duration fire cycle?

    A short-duration fire cycle will regularly prescribe fire as a tool to limit the amount of fuels that are allowed to accumulate on the forest floor. Once this approach is fully implemented, the danger of highly destructive forest fires will be greatly reduced and the health of the forest will improve because competition among trees for nutrients is reduced.

MULTIPLE-CHOICE QUESTIONS

1. Which element is *not* required in order for a forest fire to occur?

   A. heat energy   C. fuel

   B. a supply of oxygen   D. <u>high humidity</u>

2. A partially decayed fuel found on the forest floor is called:

   A. <u>duff</u>   C. peat

   B. charcoal   D. coke

3. A fire that burns fuel consisting of decayed plant material deposited beneath the ground surface is which of the following fire types?

   A. surface fire   C. <u>ground fire</u>

   B. crown fire   D. firestorm

4. A sudden dramatic increase in the intensity of a forest fire is called a:

   A. crown fire   C. <u>blowup</u>

   B. draft   D. infernal combustion

5. The greatest single natural cause of forest fires is:

   A. <u>lightning</u>   C. fireworks

   B. matches   D. campfires

6. A firestorm is caused by:

   A. two bolts of lightning striking one another

   B. <u>erratic winds that move the fire into the forest canopy where there is plenty of fuel</u>

   C. a thunderstorm

   D. ash that has drifted down from the atmosphere above the fire

7. A fire that is burning out of control is called a:

   A. prescribed burn   C. cleansing fire

   B. surface fire   D. <u>wildfire</u>

8. A fire that is intentionally set in a forest for the purpose of burning the fuel on the forest floor is called a:

   A. wildfire   C. ground fire

   B. <u>prescribed burn</u>   D. blowup

9. Weather-related factors that influence the intensity of a fire and the rate at which it spreads include all of the following *except*:

   A. <u>topography</u>   C. wind

   B. precipitation   D. humidity

10. Which term best describes the mode of action of a firefighter who shovels soil on the flames at the leading edge of a fire?

    A. fire line   C. <u>direct attack</u>

    B. incendiarism   D. indirect attack

Chapter 8

WILDLIFE AND THE FOREST

OBJECTIVES

After completing this chapter, you should be able to

- identify basic needs that are common among the plants of the forest and the wild animals living in the forest

- describe the three biomes in North America that encompass most forest environments

- distinguish differences among freshwater, temperate forest, and coniferous forest biomes

- contrast the features of a lotic habitat versus a lentic habitat

- describe the relationships known to exist between wetlands and wildlife

- list similarities and differences between biomes and ecosystems

- identify the basic needs of wild animals

- define the terms *symbiosis, mutualism, commensalism*, and *parasitism*

- explain the benefits of committing private property to a conservation easement agreement

TEACHER BACKGROUND INFORMATION

An understanding of and experience in wildlife management will help you present this unit. Obtain some good reference materials such as textbooks and wildlife publications and become familiar with the definitions and applications of the terms that are common to the discipline.

INSTRUCTIONAL STRATEGIES

Engage the expertise of a professional in wildlife management as you develop a plan for a bus tour of wildlife management areas within the local area. Most students have an interest in wildlife, and exposure to real animals will help sustain their interest. Develop a worksheet prior to the wildlife tour, and discuss the questions at the tour site. These questions should address the objectives for this chapter.

LEARNING ACTIVITIES

1. Locate a wooded area near your home or school that encompasses a water source or is designated as a wetland. A land management professional with a federal or state agency such as The Bureau of Land Management or the U.S. Forest Service should be able to help locate such an area. Observe the area at different times of the day and list the kinds of birds and animals that use it.

2. Develop a bulletin board display to illustrate the three biomes that make up forest environments in North America. Identify the plant and wildlife species likely to be found in each. Request permission to present your findings to an elementary school class. Prepare an activity such as coloring a handout of a wild animal. Explain to the class what features of the environment make it possible for each animal to live there.

3. Conduct a poster contest illustrating the basic needs of wild animals. Invite a biology teacher or local conservation officer to judge the competition. Display the posters in the school or another public location.

QUESTIONS FOR DISCUSSION AND REVIEW

Essay Questions

1. What are the similar needs of plants and animals living in a forest environment?

   Wild animals require shelter, food, and protection from predators. They require safe water, clean air, and an adequate supply of food (nutrients). Plants also require a supply of safe water, adequate nutrients from the soil, and air that is free of contaminants.

2. What are some ways that a biome is different from an ecosystem?

   A biome is a living zone wherein the environment is quite similar throughout the zone. Examples of biomes include terrestrial, freshwater, and marine biomes. The terrestrial biome includes some smaller biomes such as desert, tundra, grassland, temperate forest, and coniferous forest. An ecosystem is a smaller environment located within a biome. It consists of the plants, animals, and microorganisms that interact with one another and with the nonliving features of the environment such as water, soil, and rocks.

3. How is energy from the sun made available in the diet of a predatory animal?

   Energy from the sun is captured by plants through photosynthesis. It is stored by plant tissues as sugars, carbohydrates, and oils. Herbivores obtain energy and other nutrients by eating plants. Energy is obtained by predatory animals when they eat herbivores.

4. In what ways does water fill the basic needs of wild animals?

   Wild animals use water to control their body temperature through the process of evaporation. Water is also used to dissolve and transport the nutrients that animals eat. Indirectly, water stores atmospheric heat and stabilizes the temperature of the environments in which wild animals live.

5. What are some ways that the amount of living space affects wild animals?

Wild animals need enough living space to produce an adequate supply of food. When too many animals occupy an environment, the weaker animals must move to another location or starve to death.

   Some kinds of animals have a high tolerance for other animals of their own kind. They form herds and colonies, and when food supplies are inadequate, they tend to live or starve together. Other animals are very territorial, and the strongest animals will drive others of their own kind away.

6. How does acid precipitation affect plants and animals in forest environments?

Acid precipitation causes many kinds of plants to be damaged or killed. Either of these results reduces the food supply of those animals that depend on the affected plants for food. Predators that depend on the herbivores for food will also suffer from a short supply of nutrients. The end result over an extended period will eventually be a reduction in plant and animal populations.

7. Define parasitism and give an example of this relationship in the forest.

Parasitism is a relationship between organisms wherein one organism is benefited and the other is harmed. Examples of parasitism in the forest include the broomrapes such as squawroots, beech drops, and cancer roots.

8. In what ways does a forest function to remove pollutants from the environment?

The plants in a forest remove carbon dioxide and other gases from the atmosphere. Forest plants also filter contaminants such as heavy metals, phosphates, and nitrates from water as it moves through the water cycle.

9. What is a conservation easement?

A conservation easement is a permanent agreement between a landowner and a land trust or government agency. Participating landowners place permanent limitations on the use of the land to enhance conservation efforts and to limit development of the property. This reduces the market value of the property and also estate taxes in the event the owner wishes to transfer ownership to the heirs.

10. What kinds of duties does a wildlife conservation officer perform?

Wildlife conservation officers work with wildlife populations by performing duties such as restoring habitats, conducting population surveys, coordinating wildlife and hunting education programs, working to protect threatened or endangered species, tracking and observing animals to improve management plans, controlling nuisance animals, developing fish/wildlife management plans, and providing enforcement for laws affecting fish and wildlife.

MULTIPLE-CHOICE QUESTIONS

1. Which kind of habitat does a fast-flowing river represent?

   A. lentic   C. temperate

   B. lotic   D. coniferous

2. Which need is *not* considered a basic need of wild animals?

   A. food   C. space

   B. shelter   D. climate

3. Which biome is *not* considered to be available in a forest environment?

   A. saltwater   C. freshwater

   B. coniferous forest   D. temperate forest

4. Approximately what percent of the tissue of living plants and animals is composed of water?

   A. 15%   C. 52%

   B. 35%   D. <u>70%</u>

5. An instance in which two kinds of animals live together in a close association that is beneficial to both within the same environment is known as:

   A. commensalisms   C. compensatory

   B. parasitism   D. <u>mutualism</u>

6. A relationship between two animals that benefits one animal and harms the other is called:

   A. commensalisms   C. compensatory

   B. <u>parasitism</u>   D. mutualism

Chapter 9

WATER QUALITY AND WATERSHED MANAGEMENT

OBJECTIVES

After completing this chapter, you should be able to

- define terms that relate to watersheds
- explain the relationships that exist between healthy watersheds and sustainable forests
- illustrate the basic elements of the hydrologic cycle
- distinguish between infiltration and permeability
- explain the difference between point-source pollution and non–point source pollution
- describe the problems associated with sediment as it relates to water pollution
- list some ways that acid precipitation affects forests
- relate the importance of water quality standards to the improvement of water quality
- give some examples of beneficial uses or use classes for water resources
- list some procedures that have been developed for the purpose of reclaiming polluted water

TEACHER BACKGROUND INFORMATION

A good understanding of the hydrologic cycle and management of healthy watersheds will be critical in helping students appreciate other important relationships. Concentrate on defining the relationships between and among health of watersheds, sustainable forests, water pollution, and water quality standards.

INSTRUCTIONAL STRATEGIES

Request the use of training films or videos from state and federal agencies that teach the concepts of water quality and watershed management. Use appropriate segments to illustrate the critical factors that affect our watersheds and water quality. Among the effective strategies for teaching this subject are research papers and public speaking. Consider either or both of these approaches to teaching this topic.

LEARNING ACTIVITIES

1. Divide the class into teams of two to three students. Assign each team to conduct an Internet search to find examples of damage to forests as a result of acid precipitation. Print materials from Internet sources or publications and have each team prepare an informative poster. Each team should be allotted 2 to 3 minutes to report, using their poster to illustrate what they found.

2. Compile a list of impaired waters within your state. Identify the factors that are responsible for the listing. Research what is being done to resolve the problems that led to the listing. Invite representatives of government agencies, such as the EPA or the state department of water resources, to help in assembling the list.

QUESTIONS FOR DISCUSSION AND REVIEW

Essay Questions

1. Define the following terms: surface runoff, catchment, water tower, and headwater system.

   Surface runoff: water that flows directly to streams and rivers across the soil surface

   Catchment: same as watershed; all of the land surfaces from which water flows into a particular stream

   Water tower: a catchment basin in a forested mountain area

   Headwater system: catchment areas or watersheds located at the top end of a stream or river system

2. Why are healthy watersheds and sustainable forests both needed in order to have a dependable supply of high-quality water?

   Healthy watersheds reduce or prevent pollution of surface waters as a result of silt. In most instances, the soil is held in place by the roots of trees and other forest plants. A healthy forested watershed improves infiltration of precipitation into the soil, later to be released from springs at lower elevations. Healthy watersheds and sustainable forests are codependent. One cannot exist without the other.

3. What are the steps that occur as part of the hydrologic or water cycle?

   The cycle begins with condensation of water vapor in the atmosphere, forming small water droplets and clouds. Winds carry clouds and moist air to the interior of the continent. The clouds lose precipitation as they move inland. Rain and snow infiltrate into the soil, recharging the aquifer with fresh water. Gravity draws the groundwater down to lower elevations, where it emerges from springs and artesian wells. Some of the water flows back to the ocean, and some is evaporated from the surfaces of plant leaves and soil particles. The energy of the sun heats all surface water, causing some of it to evaporate. The resulting water vapor starts the hydrologic cycle all over again.

4. What are some examples of point-source and non–point source pollution?

   Examples of point-source pollution include areas where surface water flows across bare soil, where it picks up sediment. Pollution sometimes occurs when underground fuel storage tanks leak fuel into the groundwater supply. Spills of hazardous chemicals are known to come into contact with water, which carries the pollutant into a stream or domestic wells.

Examples of non–point source pollution often include sediment or dissolved nutrients such as nitrates and phosphates. It is often difficult or even impossible to identify the exact locations where such pollutants entered the water supply.

5. Why is sediment considered to be the most significant pollutant to come from forested areas?

Sediment occurs more often in polluted waters than any other single pollutant. Most pollution of water as a result of sediment occurs in areas where the soil surface has been disturbed. Mines, construction sites, logging areas, and plowed surfaces contribute to the presence of sediment in surface water.

6. What can be done to reduce sediment pollution in forest waters?

One of the best strategies to reduce sediment in surface waters is to create forest buffer zones alongside flowing streams and rivers. Buffer zones trap sediment as it flows over the surface, causing the sediment to be deposited in forested areas. Dirt roads contribute to sediment pollution by channeling flowing water across the bare surface and eroding the soil. Proper road construction directs water flow into trees and away from streams. Plastic barriers are also used extensively near construction sites to slow the flow rate of water. This allows soil particles to settle out of the flowing water before it enters a stream, lake, river, or pond.

7. What are some forest practices that can be used to reduce the erosion of forest soils?

Forest practices that reduce the erosion of forest soils include:

• Reforestation following a fire or timber harvest
• Restriction of clear-cut harvesting to relatively small zones interspersed with forested zones
• The use of forest buffer zones near riparian areas
• Engineering of roads to control water flows over or along the road surface

8. How are forests affected by acid precipitation?

Leaves and needles of trees may decrease in size or turn yellow. Some may even die and fall off the trees. Other damage includes loss of fine roots and diebacks of twigs and crowns of trees. Production of timber products usually declines in the long term. In some instances, entire forests have been known to die. Acid precipitation tends to cause a shortage of the soluble calcium available to plants.

9. How are water quality standards used to bring about improvements in water quality?

The creation of water quality standards provides a benchmark that helps define water quality. Failure to meet water quality standards is often accompanied by penalties to the responsible party, agency, or state. When water quality standards are enforced, greater attention is paid to eliminating soil and forest management practices that contribute to water pollution. Federal funds may be withheld from states that do not meet water quality standards. Such penalties often result in a concerted effort to identify causes and sources of water pollution.

10. What are some methods used to reclaim polluted water?

Reclamation of polluted water may require professional help in applying scientific practices to water management. Such individuals are often aware of practices implemented in the past to reduce or eliminate pollution.

Much of the effort to eliminate water pollution has centered on the cleanup of the environments around abandoned mines, logging areas, and industrial sites. In addition, state and federal water quality standards have been raised in an effort to prevent additional pollutants from entering the water. In some instances, contaminated sediments have been sealed off to prevent leakage into waterways. Sometimes all of the sediment must be removed from a contaminated site and treated in a hazardous waste facility to capture or remove the pollutants.

MULTIPLE-CHOICE QUESTIONS

1. Which term means the same thing as *watershed*?

   A. water tower

   B. pump house

   C. English bathroom

   D. catchment area

2. Which term describes the process by which water vapor is changed to a liquid?

   A. condensation

   B. evaporation

   C. permeability

   D. infiltration

3. By what process does water seep into soil?

   A. condensation

   B. evaporation

   C. permeability

   D. infiltration

4. The amount of heat required to convert water from a liquid to a gas is:

   A. heat of vaporization

   B. Celsius

   C. evaporation

   D. condensation

5. A type of water contamination that can be traced back to its source is:

   A. non–point source pollution

   B. pathogenic pollution

   C. NPS

   D. point-source pollution

6. Which element is *not* a form of sediment?

   A. clay

   B. sand

   C. coliform

   D. silt

7. Which element is closely involved with acid precipitation?

   A. sulfur

   B. water tower

   C. infiltration

   D. TMDL

8. Which term is *not* an EPA term of water contaminants?

   A. organic contaminants

   B. radionuclides

   C. inorganic contaminants

   D. bioindicator

Chapter 10

THE ROLE OF GOVERNMENT IN FORESTRY

OBJECTIVES

After completing this chapter, you should be able to

- relate the early history of the roles that government played in forest management

- describe the impact of the Homestead Act on native American forests

- identify the main purpose of the Timber Culture Act of 1873

- express the two philosophies that influence forest planning in the United States

- explain why a management plan is needed for forest resources

- name the two federal departments responsible for managing most of our forest resources

- list the most important forest management agencies and explain their responsibilities

- describe how the Forest Reserve Act of 1891 changed the way forest resources were administered

- explain how the national forests, national monuments, and national parks were established

- discuss some of the differences that are likely to occur in the ways that national, state, and privately owned forests are managed

- distinguish between the multiple-use and dominant-use concepts of forest management

TEACHER BACKGROUND INFORMATION
Government plays a major role in forest management, and even privately owned forests are subject to some of the laws that restrict forest uses and management strategies. You should become familiar with key legislation that has established the regulations by which forests are managed today.

INSTRUCTIONAL STRATEGIES
Identify the amount of forest land that is located in your state (see Appendix V). How much is controlled by the national forests in your state? (Refer to Appendix IV.) Also seek information about how much forest land is managed by the state government. How do these figures compare with privately held forest lands? Discuss whether forests should be retained as public lands or converted to private forests.

LEARNING ACTIVITIES
1. Obtain a map of a forest region on which private, state, and federal ownership of the land is identified. Invite a forest manager to talk with the class about ways that laws and government regulations and policies affect the management of the forest resources for which he or she is responsible. If you do not have a forest near your community, you might do the same activity using a speaker phone or through an Internet connection.
2. Assign students to search the Internet for evidence of the two competing philosophies of forest management. Classify each document according to its orientation toward conservation or preservation of resources and discuss the findings of the class members.
3. Prepare a bulletin board display featuring national parks, national monuments, or national forests in your region.

QUESTIONS FOR DISCUSSION AND REVIEW
Essay Questions
1. What roles did the government play in forest management in the colonial period?

   The roles of government in forest management during the colonial period were not well defined, and the result was that forest resources were not managed very well. Harvests were not controlled, and the forest resources were often wasted and exploited.
2. How did the Homestead Act impact native American forests?

   The Homestead Act required the land to be developed into productive farms. This resulted in the forests being cut, piled, and often burned to prepare fields for crop production.
3. What was the main purpose of the Timber Culture Act of 1873?

   The purpose of the Timber Culture Act of 1873 was to develop forest resources in the Great Plains region where few trees existed. Homesteaders were required to plant 40 acres of trees on each 160-acre homestead.
4. State the two different forest management philosophies that influence forest planning in the United States.

   The two forest management philosophies that influence forest planning in the United States are the conservation philosophy and the preservation philosophy. Conservation allows such forest uses as timber harvests, grazing, recreation, mining, and hunting and fishing. Preservation is focused on saving the resource while allowing human activities such as camping, hiking, boating, and fishing.
5. Why is it important to develop management plans for forest resources?

   It is important to develop management plans for forest resources because failure to implement management plans usually leads to abuse of the resources. Properly implemented management plans can prevent serious, irreversible environmental damage, such as soil erosion.

6. Name two federal departments and the key agencies that are responsible for the management of most public forest lands.

The two federal departments that manage most of the forest lands in the United States are the Department of the Interior and the Department of Agriculture. Key agencies in the Department of the Interior that manage forest lands are the Bureau of Land Management, Bureau of Indian Affairs, and National Park Service. The U.S. Forest Service is an agency within the Department of Agriculture that manages most of the forests that the public owns.

7. Describe how the Forest Reserve Act of 1891 changed the way that forest resources were managed.

The Forest Reserve Act of 1891 changed the way that forest resources were managed by giving power to the president of the United States to set aside public land as forest reserves. An example of the use of this law was the establishment of the Yellowstone forest reservation that later became part of Yellowstone National Park.

8. What events led to the classification of some forest lands as national forests, national monuments, and national parks?

Events that led to the establishment of national forests include legislation that transferred forest reserves from the Department of the Interior to the Department of Agriculture in 1905. When the transfer was complete, the Division of Forestry was changed to the U.S. Forest Service, and the forest reserves became national forests.

The American Antiquities Act in 1906 allowed the president of the United States to set aside public land to preserve sites that were of historic or scientific value. These sites became national monuments.

National parks may be established from public lands by passing laws in the Congress. The National Park Service was created following the passage of the National Park Act in 1916.

9. Name one way that a national law may affect forest lands whether they are under federal, state, or private ownership.

One way that a national law may affect federal, state, and private forests is to restrict the uses of the lands for the purpose of preserving habitat for a threatened or endangered species. Land uses may also be restricted under provisions of the Clean Air Act, the Federal Water Pollution Control Act, and other legislation.

10. What are the differences between managing forest resources for multiple-use purposes in contrast with dominant-use management?

Multiple-use forest management is a concept that allows different forest resources to be used simultaneously within an area or different uses of a single resource at the same time. Dominant-use management assigns a greater priority to a particular forest use than to alternative uses.

MULTIPLE-CHOICE QUESTIONS

1. Forest management in colonial times could best be described as forest:

   A. conservation   C. dominant-use management

   B. <u>exploitation</u>   D. preservation

2. Which effect did the Homestead Act of 1862 have on forest resources?

   A. <u>forest acreage decreased</u>   C. erosion of forest soils declined

   B. forest acreage increased   D. the quality of surface water improved

3. The Timber Culture Act of 1873 required that:

   A. trees could no longer be harvested

   B. timber products could not be exported

C. timber lands could be purchased in 160-acre blocks

D. <u>forty acres of each homestead in the Great Plains region had to be planted in trees</u>

4. In general, the philosophy for management of forest resources administered by the U.S. Forest Service could be described as:

A. exploitation

C. <u>conservation</u>

B. preservation

D. dominant-use management

5. Which government organization is *not* involved in managing forest resources?

A. U.S. Fish and Wildlife Service

C. U.S. Forest Service

B. Department of Defense

D. <u>Social Security Administration</u>

6. National parks and monuments are administered by the:

A. <u>Department of the Interior</u>

C. U.S. Forest Service

B. Bureau of Indian Affairs

D. Department of Agriculture

7. Multiple-use forest management as it appears in federal laws approves each of the following forest uses with the *exception* of:

A. grazing

C. wildlife habitat

B. <u>farming for crop production</u>

D. recreation

8. The ability of a forest to produce a volume of timber products that can be repeated year after year is referred to as:

A. gross annual production

C. diminishing returns

B. low input/high yield

D. <u>sustained yield</u>

9. The USDA Cooperative Extension System employs forest resource specialists who provide information and consulting services to:

A. managers of privately owned forests

C. state forest managers

B. U.S. Forest Service

D. <u>all of these</u>

10. A forest management strategy that requires forest resources to be managed for several different purposes is known as:

A. <u>multiple use</u>

C. resource preservation

B. dominant use

D. competitive advantage

Section Four
FOREST PRODUCTS

Chapter 11

WOOD CONSTRUCTION MATERIALS

OBJECTIVES

After completing this chapter, you should be able to

- distinguish between processed woods of the hardwood and softwood varieties

- list some distinguishing characteristics that are useful in identifying woods of different species

- identify some characteristics of wood that contribute to its value for construction purposes

- identify some characteristics of wood that detract from its value for construction purposes

- describe the general process by which logs are processed into lumber

- classify the different cuts of processed wood according to their dimensions

- distinguish between lumber and timbers

- identify some characteristics of hardwoods that contribute to their usefulness

- explain the source and methods of processing wood veneers

- distinguish the differences among the different types of reconstituted boards

TEACHER BACKGROUND INFORMATION

Reconstituted wood products have taken over a major part of the building construction market in North America. Plywood is convenient to use, and it is stronger than lumber for some uses. Construction-grade plywood and products like it contain high-quality wood on one side of the board but lower-grade wood in the middle plies or layers. In this way, low-quality wood can be made into a strong and useful product.

INSTRUCTIONAL STRATEGIES

Obtain pieces of the standard sizes of lumber products named in essay question 4. Also obtain samples of each of the laminated and reconstituted wood products. Lead a classroom discussion on what each of the wood products is used for. Conduct a similar discussion using samples of softwoods and hardwoods of the most widely used species. This exercise might be used to introduce the unit, but you may want to save it for use at a later stage of instruction for this chapter.

LEARNING ACTIVITIES

1. Visit a home building supply warehouse, either individually or as a class, and report on the dimensions, grades, and prices of wood products that were available for home construction. Assign students to small work groups, and give them some log dimensions (debarked). Have each group calculate the highest possible retail value that could be obtained from the log based on the lumber dimensions and prices observed at the warehouse.

2. Obtain some samples of reconstituted wood products and allow the students to examine them. Assign students to small work groups and have each group do an in-depth written and oral report on the types of materials used to make the product. Have each group explain the probable process by which the product was manufactured.

QUESTIONS FOR DISCUSSION AND REVIEW

Essay Questions

1. What differences exist between processed woods of the hardwood and softwood varieties?

   Hardwoods and softwoods have some distinctly different characteristics. Pores are present in hardwoods but not in softwoods. Resin ducts are present in softwoods but not in hardwoods. Hardwoods are higher in density (heavier) than softwoods.

2. What properties of wood make it suitable as a material for construction purposes? What properties detract from its usefulness?

   Wood properties that contribute to its suitability for construction purposes include durability, toughness, tensile strength, aesthetic properties, insulating qualities, and flexibility. Properties that detract from its usefulness include its tendency to shrink and swell, the fact that its fibers cannot be welded and it is not easily molded to new shapes, its lack of uniformity, and the fact that it breaks down over time.

3. Describe the general process by which logs are processed into lumber.

   Logs are generally processed into lumber in the following steps: debarking (removal of bark), sorting by size, X-ray scan to evaluate the highest-value cuts, slab removal by the head saw, and cutting the logs into lumber. A head saw cuts the lumber in a series of cuts; a gang saw may be used to cut a small log into lumber in a single pass. Boards move to sorting sheds on the green chain, where they are planed, trimmed, graded, sorted, and stacked for drying.

4. List the different cuts of processed wood according to standard lumber and timber dimensions.

   The different cuts of wood according to standard lumber and timber dimensions include:
   - Lumber: less than 5″ × 5″
   - Board: less than 2″ thick and greater than 4″ wide

- Plank: $1\frac{7}{8}''$ to 4″ in thickness and greater than 11″ wide
- Timber: greater than 5″ × 5″
- Beam: greater than 8″ × 8″

5. List the duties performed by a person who chooses a career as a lumber grader.

A lumber grader inspects lumber products for defects as they move across the conveyor system. The grader is looking for decay, splits, milling defects, knots, and stains. He or she assigns a grade to the product, and the lumber is measured to ensure standard dimensions.

6. What force causes green lumber to warp when it is not properly stacked as it cures?

Freshly cut lumber (green lumber) contains much of the water that was present in the live tree. Once it has been processed, the water begins to evaporate from the lumber. As water is lost, it is lost at different rates from different sides and sections of the board. This causes the wood to shrink unevenly, causing boards to become twisted out of shape or warped.

7. What are some ways that structural wood products are used?

Structural wood products are used to construct building frames, bridges, railroads, utility poles and cross members, and pilings for docks and bridges, as well as to shore up mine shafts.

8. What are some major uses for which hardwoods are best adapted?

Major uses for which hardwoods are best adapted include floors, furniture, doors, chairs, veneers, paneling, and industrial pallets.

9. Explain what veneer is and how it is obtained from logs.

Veneer is a thin layer of wood (¼″ or less in thickness) that is sliced from a high-quality wood flitch or peeled from a high-quality log or block using a specialized lathe.

10. What is the process by which plywood is manufactured?

Plywood is manufactured by laminating several layers of wood veneer together using glue, heat, and pressure to form the plywood product. The long length of wood fibers are aligned across one another in each layer (cross-banded) to add strength.

11. How are reconstituted boards such as particleboard, fiberboard, and hardboard made?

Reconstituted boards are made by using wood shavings and/or sawdust mixed with adhesives. This material is placed between steel plates, where it is subjected to heat and pressure until the glue sets. Particleboard, fiberboard, and hardboard are all formed using variations of this process.

12. How are plywood, particleboard, fiberboard, and hardboard different from one another?

Plywood consists of wood in which the wood fibers in each layer have been left intact, as they were in the tree. Cross-banding takes advantage of the tensile strength of wood and makes plywood stronger than boards across its length and breadth. Particleboard tends to exhibit similar characteristics to plywood because of the random arrangement of the wood chips from which it is made. It does not have a smooth exterior, however, that is capable of being finished to a smooth surface. Hardboards are smooth-surfaced, but they lack tensile strength because of the small size of the sawdust particles from which they are made.

MULTIPLE-CHOICE QUESTIONS

1. What is a distinguishing visual feature of hardwoods?

A. <u>pores</u>   C. density

B. resin ducts   D. odor

2. What is a distinguishing feature of a softwood?

   A. pores   C. <u>resin ducts</u>

   B. color   D. heartwood

3. The strength of wood along the long axis of a tree is called:

   A. ductile strength   C. toughness

   B. density   D. <u>tensile strength</u>

4. A saw that makes two or more cuts at the same time is called a:

   A. <u>gang saw</u>   C. head rig

   B. band saw   D. circular saw

5. A log from which only the slabs have been removed is called a:

   A. timber   C. beam

   B. <u>cant</u>   D. block

6. A timber with a cross-section measurement of 8″ × 8″ or greater is called a:

   A. cant   C. <u>beam</u>

   B. block   D. flitch

7. A source of power used in lumber mills that is provided by compressed air is called:

   A. hydraulic power   C. wind power

   B. <u>pneumatic power</u>   D. steam power

8. A condition that occurs as green lumber dries unevenly, resulting in distorted shapes in the wood, is known as:

   A. lamination   C. symmetry

   B. asymmetry   D. <u>warp</u>

9. What is a thin sheet of wood peeled from a block called?

   A. <u>veneer</u>   C. particleboard

   B. plywood   D. reconstituted wood

10. The practice of aligning long wood particles and fibers across one another in the manufacture of reconstituted wood is called:

    A. articulation   C. integration

    B. <u>cross-banding</u>   D. cross-bonding

11. Another name by which hardboard is well known is:

    A. particleboard   C. <u>Masonite</u>

    B. waferboard   D. fiberboard

12. A reconstituted wood product in which wood shavings are a major component is called:

    A. waferboard   C. hardboard

    B. fiberboard   D. <u>particleboard</u>

Calculations

Refer to the **Standard Lumber Sizes** discussed in this chapter and determine the maximum number of boards that can be expected from each log (use rough-cut dimensions). **Helpful Hints: Assume that the lumber pieces are the same length as the log. Draw a view of the small end of the log to scale. Cut some paper pieces of the lumber dimensions to scale and fit them into the end drawing. Remember that the saw will require ¼″ for each cut, and each cut extends completely through the log. We also assume that the order in which the cuts are made is correct for maximizing the number of designated lumber sizes.**

1. Determine the number of 2 × 4s that can be obtained from each log:

    A. Log dimensions: large end, 6″ diameter (d); small end, 4.9″ d

       Note: The small end of the log is the limiting factor. This log will just barely provide two 2 × 4s, assuming it is straight and without defects.

    B. Log dimensions: large end, 12″ d; small end, 10.6″ d

       Six 2 × 4s can be cut from this log.

2. Determine the maximum number of 2 × 6s and 1 × 4s that can be cut from each log: **Hint: Assume that this is a combination of 2 × 6s and 1 × 4s.**

    A. Log dimensions: large end, 10″ d; small end, 9.2″ d

       Three 2 × 6s and four 1 × 4s

    B. Log dimensions: large end, 9.8″ d; small end, 8.5″ d

       Two 2 × 6s and four 1 × 4s

3. Determine the maximum number of 1″ thick boards of various widths (with emphasis on wide boards) that can be cut from each log:

    A. Log dimensions: large end, 22″ d; small end, 20″ d

       16, 1 × 12s; 2, 1 × 8s; 8, 1 × 6s; 4, 1 × 4s; 4, 1 × 2s

    B. Log dimensions: large end, 34″ d; small end, 32″ d

       46, 1 × 12s; 2, 1 × 10s; 2, 1 × 8s; 12, 1 × 6s; 6, 1 × 4s; 4, 1 × 2s

---

Disclaimer: This is a good visual activity for students, but in actual practice, the number of cuts required to maximize lumber recovery would be likely to slow down the movement of logs through the mill. The loss of efficiency may be more costly than the value of recovered lumber.

Chapter 12

SPECIALTY FOREST PRODUCTS

OBJECTIVES

After completing this chapter, you should be able to

- distinguish between a monosaccharide and a polysaccharide

- explain the significance of cellulose to the fiber and paper industries

- describe the different methods used to convert wood fiber to pulp

- suggest reasons why the mechanical pulping method is widely used in the paper and fiber industries

- distinguish the differences between the bleaching and brightening processes

- explain why different grades of wood pulp are sometimes blended together

- define the process by which wood is converted to ethanol

- identify products obtained by destructive distillation of wood

- name the different types of products extracted from wood using solvents and explain the processes by which they are obtained

- discuss the importance of biomass as a fuel for generating electrical power

TEACHER BACKGROUND INFORMATION

Wood products include much more than the lumber used for construction purposes. A huge industry has developed around wood pulp, which is used to make cardboard and other paper products. The most common products of wood are discussed in this chapter, and it is very evident that many useful products are extracted from wood. Many states have commissions that promote and market forest products. These commissions are excellent sources of educational materials and classroom speakers.

INSTRUCTIONAL STRATEGIES

Organize the students into groups of four to five students and conduct a competitive activity to see which group can assemble the most diverse group of products obtained from wood. This exercise will help students understand that forests are important sources of many products used in their homes.

LEARNING ACTIVITIES

1. Obtain microscopes with which to observe the structure of paper and paper products. Point out the overlapping structure of the fibers, and explain to the students that chemical and ionic bonds also attract and hold the wood fibers together.

2. Collect as many products as you can find that are obtained from wood. Assign pairs of students to make and display posters illustrating how one of the products was manufactured. Give each group of students an opportunity to discuss their product with the class. Keep the collection of products together to be used in future classes.

QUESTIONS FOR DISCUSSION AND REVIEW

Essay Questions

1. What is the relationship between monosaccharides and polysaccharides?

   The relationship between monosaccharides and polysaccharides is that polysaccharides consist of chains of monosaccharides that have bonded together to form complex molecules such as starch and cellulose.

2. Why is cellulose important to the fiber and paper industries?

   Cellulose is important to the fiber and paper industries because it is the raw material from which cardboard and paper products are made.

3. What is the purpose of adding chemical pulp to mechanical pulp in the manufacture of newsprint?

   The purpose of adding chemical pulp to mechanical pulp in the manufacture of newsprint is to increase the strength of the paper.

4. Name the different pulping processes, and compare the methods that are used by each of them.

   The different pulping processes and the methods by which each is manufactured are:

   • Mechanical pulping: A stone grinder or disk refiner is used to separate the wood fibers as it grinds up pulpwood bolts or wood chips.

   • Semichemical pulping: Wood is exposed to a mild chemical treatment to partially separate the fibers before it is processed through a disk refiner.

   • Chemical pulping: Chemicals are used to dissolve the lignin component of wood in a large container called a digester. Heat and pressure are applied, and lignin is dissolved in water and removed.

   • Hydrapulping: Recycled paper is mechanically reduced to pulp in a hydrapulper.

5. Explain the bleaching and brightening processes, and describe the differences between them.

Bleaching is a process by which pulps are treated with chemicals to remove dark coloring, resulting in different degrees of whiteness in the final paper product. Brightening is a process used to alter the lignin component of pulp, making a lighter-colored compound.

6. Describe the process by which wood is converted to ethanol.

The process by which wood is converted to simple sugars is called saccharification. Fermentation is the process by which the simple sugars are converted to a fuel-grade alcohol called ethanol.

7. What are the end products of the process called *destructive distillation*?

The end products of the process of destructive distillation are charcoal and volatile gases.

8. What are some extracted products, and how are they obtained from wood?

Extracted products from wood include organic solvents such as wood resins (for example, turpentine), rosin, fat, fatty acids, tannins, and lignins. All of these products are removed from wood by dissolving them in organic solvents or water.

9. List some commercial products obtained from naval stores.

Commercial products obtained from naval stores include turpentine, resin, synthetic pine oil, insecticides, flavor and fragrance chemicals, tannins, and rosins.

10. What is *biomass*, and how is it used commercially?

Biomass is wood that is chipped and dried to be burned as a source of energy. It consists of leaves, branches, stems, and roots, as well as recycled wood and waste materials from lumber processing mills.

MULTIPLE-CHOICE QUESTIONS

1. What is another name for a simple sugar that forms in long chains to make cellulose?

   A. monomer
   B. lignin
   C. hemicellulose
   D. polysaccharide

2. A pulping process in which wood fibers are separated from each other by grinding or abrasion is called:

   A. chemical pulping
   B. hydrapulping
   C. mechanical pulping
   D. semichemical pulping

3. A pulping process in which wood fibers are separated by dissolving the lignin that cements them together is called:

   A. hydrapulping
   B. chemical pulping
   C. mechanical pulping
   D. disk refining

4. What machine is used to reduce recycled paper to pulp?

   A. Fourdrinier
   B. stone grinder
   C. disk refiner
   D. hydrapulper

5. A process that changes the lignin in paper pulp to a compound that is lighter in color is known as:

   A. bleaching
   B. brightening
   C. coloring
   D. blending

6. Which products does *not* come from cellulose xanthate?

   A. naval stores
   B. photographic film
   C. cellophane
   D. rayon

7. A fuel composed entirely of the product obtained by converting cellulose to simple sugars and fermenting them to form alcohol is called:

    A. methanol   C. gasohol

    B. charcoal   D. <u>ethanol</u>

8. What is the process that produces oil from wood called?

    A. destructive distillation   C. <u>thermochemical liquefaction</u>

    B. saccharification   D. fermentation

9. Oleoresin is a product obtained by collecting the sap from trees. Which materials contain oleoresin?

    A. wood naval stores   C. sulfate naval stores

    B. <u>gum naval stores</u>   D. food grade margarine

10. Which material produces the hottest flame when it is burned?

    A. hardwood   C. <u>charcoal</u>

    B. softwood   D. biomass

Chapter 13

PLANTATION PRODUCTS AND PRACTICES

After completing this chapter, you should be able to

- define *monoculture* as it relates to forestry
- explain the cultural practices involved in the production of containerized seedlings
- describe how cuttings are used to produce trees for transplanting
- list some silviculture practices used in the management of a Christmas tree plantation

- distinguish between bare-root stock and containerized seedlings
- explain the beneficial effects of pruning on lumber quality
- evaluate the production of biomass as a plantation crop in contrast with biomass production in most forest environments

TEACHER BACKGROUND INFORMATION

Production of trees on large, intensely managed plantations is becoming commonplace in many regions of North America. The forest products produced in a plantation setting usually yield at higher levels than those from natural forest environments. This is because many of the factors that reduce production yields in naturally occurring forests are controlled in plantation settings. Pest-resistant, hybrid trees of a single high-yielding variety are often used to replace the native trees when forests are converted to intense management systems in tree plantations.

INSTRUCTIONAL STRATEGIES

Locate a Christmas tree plantation and either take a field trip to the tree plantation or invite the manager of the plantation to speak to the class about proven management practices. Remember to prepare a worksheet to guide the students in their questions and to help them organize their written reports after they return from the field trip. Let the field trip be the introduction to this chapter.

LEARNING ACTIVITIES

1. Obtain some nursery supplies and select some seeds for trees that are adapted to your area. Instruct students on the proper planting procedures, and have each student plant a small tray of seeds. Have the students identify their trays, and make each student responsible for caring for his or her own plants as they sprout and grow. Keep the plantings damp and maintain them in a warm place as they germinate. Once they have emerged, place the plantings in a greenhouse or near a window to allow exposure to light. Allow the students to take their seedlings home, or find a planting location near the school when the trees are big enough to be transplanted outside.

2. Take a field trip or assign students to visit a local nursery. Make a list of the trees available in landscape and shade tree varieties. Ask the customer service representative to explain how these trees should be transplanted for best results.

QUESTIONS FOR DISCUSSION AND REVIEW

Essay Questions

1. Explain why plantation forest plantings tend to be monocultures.

   Plantation plantings tend to be monocultures (a single species) because the same production practices can be applied to the entire stand of trees. A monoculture in a plantation is a crop that matures uniformly and responds well to intensive production practices.

2. What steps are involved in the production of containerized seedlings?

   Containerized seedlings are produced in individual packets of soil inside greenhouses, where they have adequate water and protection from excessive heat and cold temperatures. Seedlings grow under these controlled conditions until they are big enough to be successfully transplanted in the forest.

3. How are cuttings used to produce young trees for transplanting?

   Cuttings usually consist of tips of tree branches placed in a soil medium to promote the development of roots. After the roots form, they remain in greenhouses or cold frames until they mature sufficiently to be transplanted.

4. What are two silviculture practices used in the production of Christmas trees that cause the trees' branches and foliage to increase in density?

   Silviculture practices that are used in the production of Christmas trees to cause the branches and foliage to increase in density include shearing them and pinching off the central leader of the tree. This causes additional branches to be produced.

5. Explain how pruning the lower branches from the tree's stem usually results in improved lumber quality.

Pruning the lower branches from a tree improves the lumber because it allows a uniform shell of wood to be deposited that is free of knots. Knots are the extensions of branches into the trunk of the tree.

6. Distinguish between containerized seedlings and bare-root stock.

Containerized seedlings are young trees that have a ball of soil attached to their roots at the time they are removed from the containers during transplanting. Bare-root stock consists of seedlings that have been removed from the soil. The roots of these seedlings must be protected from freezing, and they must be kept moist until they are transplanted.

7. How is a biomass planting under plantation conditions superior to biomass production in a natural forest environment?

The advantage of producing biomass plantings in plantations in comparison with a forest environment is that much higher production is possible in the plantations. This is a result of proper spacing of the trees to allow for reduction of competition among trees, weed control, protection from insects and diseases, additions of fertilizer, use of fast-growing hybrid tree varieties, and supplemental irrigation.

MULTIPLE-CHOICE QUESTIONS

1. What is a forest planting consisting of a single variety of tree called?

   A. plantation   C. nursery

   B. <u>monoculture</u>   D. silviculture

2. A young tree generated from seed in a container filled with soil is a:

   A. sapling   C. <u>containerized seedling</u>

   B. cutting   D. twig

3. What is a young tree generated by vegetative reproduction called?

   A. sapling   C. containerized seedling

   B. <u>cutting</u>   D. twig

4. The shape and density of the branches and foliage of a Christmas tree are improved by which practice?

   A. <u>shearing</u>   C. flocking

   B. cutting   D. culturing

5. Lumber quality can be improved by eliminating knots from the main stem of a growing tree using a cultural practice called:

   A. cutting   C. shearing

   B. culturing   D. <u>pruning</u>

6. A young tree that has had its roots removed from the soil in preparation for planting is known as:

   A. a cutting   C. a sapling

   B. <u>bare-root stock</u> D. a containerized seedling

7. What is a dense plantation planting of fast-growing trees for energy production called?

   A. <u>biomass</u>   C. pulpwood

   B. prune production   D. jungle

8. Nursery stock consists of:

    A. young cattle or sheep that are grazed in nurseries to control weeds
    B. trees that are maintained in nurseries from which rootstock is obtained for grafting
    C. a product obtained from tree sap that is used as a soup base for human consumption
    D. <u>all of the trees and other plant materials that are maintained for sale by nurseries</u>

Section Five

NEW DIRECTIONS AND TECHNOLOGIES IN FORESTRY

Chapter 14

URBAN FORESTRY

After completing this chapter, you should be able to

- appraise career opportunities in the emerging field of urban forestry

- define the roles of trees in urban settings

- identify factors that should be considered in selecting trees for urban uses

- explain how a zone map should be used to guide tree selection

- name three basic functions of soil

- describe the relationship between soil characteristics and root development in trees

- evaluate the use of a tensiometer as a water management tool

- explain why it is important to prune trees

- describe a systematic approach to diagnosing problems in trees

- explain how cables and other hardware items are used to stabilize and repair damaged trees

- analyze the differences between the Plant Health Care (PHC) system for managing trees and traditional methods of management

TEACHER BACKGROUND INFORMATION

Most major cities now employ an urban forester who manages the care of the trees located on public property. Many of these offices also distribute literature that deals with urban forestry issues and problems. You would do well to engage urban foresters and their resources in teaching this chapter. If an urban forester is not available, a horticulture or landscape specialist or extension educator may be able to provide the same service.

INSTRUCTIONAL STRATEGIES

Gather samples of leaves from as many species of trees as you can find in your community. Identify the species and display the plant materials in the classroom as you begin this chapter. Challenge students to find each of the samples you have displayed, and offer extra credit for samples that are found in addition to those in the classroom display. Teach the students to prepare leaf samples by pressing, drying, and laminating their collections.

LEARNING ACTIVITIES

1. Develop an arboretum at or near the school consisting of the varieties of trees and woody shrubs adapted to your local area. Make the arboretum available to community groups for educational purposes. This is a long-term school project that will require special care and management.

2. Take a field trip to a public park and record the different varieties of trees and shrubs found there. Observe the trees carefully to detect any problems that may be present, and consider ways to improve the health of the trees and shrubs you observe. Have the class members write field reports that list the woody plants and trees, along with observations about the health and condition of the trees.

QUESTIONS FOR DISCUSSION AND REVIEW

Essay Questions

1. What career opportunities exist in urban forestry, and what training is required for those careers?

   Urban forestry is a relatively new career field. A university degree in forestry, horticulture, or a similar field with a strong biological science component is usually required. Many of the same principles that apply to forest management are also important in caring for trees in a populated area. Career opportunities are available in the management of trees and shrubs in city parks, along the streets, and in landscape settings.

2. What roles do trees play in cities and towns?

   Trees are important in cities and towns because they provide beauty in landscapes that would otherwise consist mostly of buildings. They cool the environment with their shade, and they provide protection from wind.

3. List some factors that should be considered in selecting a tree for planting in a town or city.

   A tree selected for planting in a town or city should have a wide range of tolerances for environmental factors such as paved surfaces that cover much of the root zone, compacted soil, and pollutants in air and water.

4. How is a zone map used in the selection of trees?

   A zone map identifies restrictive growing conditions such as altitude, climate, temperature, and availability of moisture, all of which affect the ability of plants to survive. Trees should be selected that are hardy enough to tolerate the conditions identified for the area on a zone map.

5. What are the three basic functions of soil?

The three basic functions of soil are to (1) provide support for the plant through the network of roots that spreads in the soil, (2) act as a reservoir for air and water close to the roots of the tree, and (3) provide minerals and organic nutrients to the tree.

6. How does the depth of the soil profile affect the root development of trees?

Shallow soil depth or soil that has a shallow restrictive layer prevents the roots of trees from penetrating deep enough to provide support for a mature tree. Most trees will become stunted and weak when the roots are restricted.

7. What function does a tensiometer perform that makes it useful as a water management tool?

A tensiometer is an instrument that measures the amount of moisture in soil. The ceramic tip of the tensiometer is placed at an appropriate depth in the root zone. As the soil at that depth dries out, water is lost from the instrument and vacuum pressure is created. The dial of the tensiometer provides a reading that indicates when irrigation is needed. Deep penetration of water promotes deep penetration of tree roots.

8. What useful purposes are accomplished by pruning trees?

The health of the tree can be protected by removing weak or diseased branches as they are noticed. In many cases, a diseased or damaged tree can be saved if the infection is removed before it moves throughout the entire tree. Pruning is also necessary when the tree is young in order to shape it by selecting strong, well-placed branches as the main branches. Branches that compete with the central leader should also be removed.

9. List several important considerations that should be made in diagnosing a problem with a tree.

The following questions should be asked:

- What species of trees are affected? Some insects and diseases affect only specific kinds of trees.
- What damage patterns exist? If all species are affected equally, consider pollution or weather damage as possible causes.
- What site conditions appear likely to contribute to the problem? Perhaps a restrictive soil layer exists, or the trees are not getting enough water.
- What changes have occurred at the site? Consider everything that has been done at the site.
- What visible changes have occurred in the foliage of the tree? Damage or visible changes in leaf color are indicators that can be used to discover the primary cause of the problem.
- What symptoms are visible on the branches and trunk of the tree? Look for visible signs of insects, lawnmower damage, wire-girdling, or other unusual damage.
- Are there any signs of root damage? New construction may have severed some of the root system, or the soil may be waterlogged.

10. Describe some methods for repairing some of the more common types of damage that occur to trees.

Weak branches can be supported by triangular cabling to help support their weight and hold them in position. Simple splits in the tree trunk where two branches separate into two leaders can be repaired by reinforcing the tree with bolts that hold the trunk intact in the area of the split. This is done by drilling a hole through the tree, inserting the bolt with washers, and tightening the nuts snugly on the bolt. Branches can be held apart with spacers to change the growth pattern or pulled closer together. Live bark can be grafted over an area to reestablish the flow of nutrients where a tree has recently been girdled. Hollow trees can be reinforced by filling the cavity with concrete.

11. How is the PHC system for managing the health of trees superior in some ways to traditional methods of management?

The PHC system for managing tree health is superior in some ways to traditional methods of management because it focuses on the health of the tree instead of on the tree's problems. Integrated pest management is used with many approaches to controlling damaging pests.

MULTIPLE-CHOICE QUESTIONS

1. A career that involves caring for trees in cities and towns is:
   A. extension forester   C. <u>urban forester</u>
   B. regional forester   D. district forester

2. Which tree problem is most associated with urban environments?
   A. <u>compacted soil</u>   C. insect damage
   B. disease infestations   D. weather damage

3. Which term is directly related to the hardiness of a tree?
   A. form   C. shape
   B. <u>zone map</u>   D. growth pattern

4. The most important factor to tree survival following transplanting is:
   A. soil fertility   C. <u>timely water applications</u>
   B. pruning   D. favorable temperature

5. A tensiometer is an instrument that is used to:
   A. calculate the strength of a tree
   B. measure purity of irrigation water by testing surface tension of water
   C. estimate the flexibility of tree limbs
   D. <u>measure the water content of soil</u>

6. Most problems with urban trees are the result of:
   A. <u>abiotic agents</u>   C. biotic agents
   B. insects   D. disease

7. The practice of cabling a tree is performed for what purpose?
   A. removing a dangerous tree that is damaged or diseased
   B. <u>stabilizing branches of the tree that have been damaged</u>
   C. repairing a tree crotch between two leaders
   D. controlling the direction in which the tree falls when it is cut down

8. A concept for controlling harmful insects or other pests while providing some protection for useful organisms is called:
   A. PHC   C. <u>IPM</u>
   B. GPS   D. GIS

9. A management system for trees that focuses on the health of the trees instead of on pests and diseases is called:
   A. <u>PHC</u>   C. IPM
   B. GPS   D. GIS

10. An advantage of controlling insects by engineering new tree varieties with genetic insect resistance is that:

    A. the wood of engineered trees is stronger than other wood

    B. this is a mechanical method of insect control

    C. all of the insects become susceptible to the same insecticides

    D. <u>the only insects affected are those that attempt to feed on the tree</u>

Chapter 15

SPACE-AGE FOREST TECHNOLOGIES

TEACHER BACKGROUND INFORMATION

Computers and other space-age technologies are widely used in the forest industry today. They make it possible to interpret data and to apply those interpretations in forest management plans. Prior to teaching this chapter, a phone call to a national forest headquarters office would allow the teacher to learn how computers are being used. This would give the teacher a perspective on the extent to which computers and other new technologies are being used in the forest and wood products industries.

INSTRUCTIONAL STRATEGIES

Obtain a GPS receiver from a local hunting or sporting goods store and demonstrate its use in locating positions. Take readings from two or three separate locations and plot them on a map. Some of the locations might be the homes of students in the class or prominent landmarks in the region. These readings should be taken ahead of time, but the teacher and students should take the readings near the school. To demonstrate the accuracy of the instrument, the coordinates of a hidden box of candy bars could be given to the class, and the students could use the GPS instrument to locate the treasure. The reward of doing well could be dividing up the treasure and eating it.

LEARNING ACTIVITIES

1. Plan a field trip to a government agency field office responsible for management of forest resources. Request demonstrations of computer technology to include the use of cruising computers and some computer models that the agency uses. Obtain the names of some of the computer software packages and forest modeling materials that are used.

2. Assign the class to work together in groups of two or three students to design a poster showing ways in which computers are now used or may be used in the future to manage forest resources.

QUESTIONS FOR DISCUSSION AND REVIEW

Essay Questions

1. What is a computer model, and how is it used in forestry?

   A computer model consists of advanced programming available as a software package used to imitate or simulate an actual system such as a forest. For example, a forest management model should be able to initiate the responses of the entire forest to changes in forest conditions.

2. Identify some ways that computer networks can be used in the forest industry.

   Computer networks are of value to the forest industry because a network allows entire forest management teams to have access to the same data and software. They can share files stored in a system-wide database.

3. How are computer systems used in the forest industry for financial management? Processing? Forest inventories?

   The forest industry is able to use computers to control many of the machines that are used in manufacturing wood products. This helps to maintain uniformity. Forest inventories are vast in size, and the computer makes it possible to update them as growth occurs and as timber harvests reduce them. Computer software is available to manage financial records and transactions related to the forestry enterprise. Financial reports that are generated are useful in negotiations with financial institutions such as banks and other lending agencies to obtain lines of credit and to manage cash flow. Equipment inventories can also be managed with the computer.

4. Speculate on some new ways that computers may be used to manage forest resources in the future.

Future use of computers may include the use of the earth resource satellites to monitor every forest in the world. The computer might be able to interpret satellite data to detect environmental pollution before it becomes serious enough to damage the trees. Perhaps this technology will even become advanced enough to monitor human activity in forest habitats known to be populated by threatened and endangered species of organisms.

Note: Any answer that shows some imagination in the ways that computers might be used in forest management would be acceptable.

5. Describe some ways in which military and space technologies are being adapted to new uses in forestry.

Military and space technologies include the geographic information system (GIS) and the global positioning system (GPS). The GIS combines data obtained by remote sensing technologies with computer mapping technologies. Each forest map is divided into sectors, and information such as soil type, harvest yields, forest types, and insect infestations is recorded as a management factor in each sector to which it applies. The GIS makes it possible to consider the forest management needs of each sector of the forest independently.

The GPS is capable of locating the exact location in a forest from which field data are gathered. As timber cruisers enter forest data into handheld computers, the GPS can identify the exact locations from which the data were generated. These new technologies are likely to become key management tools as environmental accountability becomes more important to forest managers.

6. How may new technologies be used to increase the production of forests?

New technologies will contribute to the production increases of forests because they will improve our ability to manage forests no matter how remote or isolated they may be. Intensive management of trees is becoming more of a reality as technology advances. As our ability to gather critical forest management data improves, we will be able to design management plans that will focus on preventing forest problems, instead of on fixing them.

7. What are some new technologies that contribute to solutions for environmental problems in forests?

New technologies that will have a big impact on resolving environmental problems include GPS, GIS, and remote-sensing technologies. These tools will make it possible to discover and monitor forest environments using computers. They may even become sophisticated enough to measure the results of efforts to fix environmental problems.

MULTIPLE-CHOICE QUESTIONS

1. A computer network that provides user access to the information contained in the databases of many smaller computer networks throughout the world is known as:

   A. <u>Internet</u>   C. database

   B. modem   D. software

2. Which criterion is *not* considered in a computer-generated forest growth model?

   A. <u>machine costs</u>

   B. competition with other trees

   C. size or age of trees

   D. quality of the growth site or environment

3. The name of a computer program that is used to predict the behavior of a forest fire is:

   A. TREES   C. ECHO

   B. SIMAC   D. <u>BEHAVE</u>

4. Which computer program is used to determine the amount of timber that can be harvested based on forest management objectives?

   A. <u>SORAC</u>   C. SPOT

   B. GROWTH   D. BEHAVE

5. An American satellite system of the type known as earth resource satellites is called:

   A. SPOT   C. <u>Landsat</u>

   B. SIMAC   D. MAX MILLION II

6. A remote sensing technology that is used to map such information as soil and forest types is known as:

   A. GPS   C. SPOT

   B. <u>GIS</u>   D. BEHAVE

7. A satellite system used to identify exact locations in a forest is known as:

   A. <u>GPS</u>   C. GIS

   B. BEHAVE   D. SIMAC

Section Six

DENDROLOGY:

The Scientific Study of Trees

Chapter 16

PHYSIOLOGY OF TREES

TEACHER BACKGROUND INFORMATION

Many different life processes occur in the world of plants. Each is controlled by chemical reactions that occur in very specific circumstances. Plant reproduction is only one example of a physiological process. A good science textbook on plant physiology, biology, or botany will be well worth the cost as you prepare to teach this chapter.

INSTRUCTIONAL STRATEGIES

Obtain some enlarged photographs of pollen grains. They can be found in a reference encyclopedia, *National Geographic* magazine, medical publications, and other sources. Pollen grains are unique and beautifully shaped structures of many different kinds. Each different kind of pollen grain has its own unique shape. Explain to the students that as they study this chapter, they will learn how these beautiful and artistic-looking structures are created in nature.

LEARNING ACTIVITIES

1. Prepare some planting pots and obtain some fresh cuttings from the suckers of a poplar tree or other similar tree species. Make sure there is a well-developed bud on each cutting. Have the students "plant" their cuttings beneath a layer of soil and care for them for several weeks. The soil should be kept damp, and the pots should be placed in warm locations for best results. Observe the plantings daily and record the results.

2. Obtain several different kinds of medium to large flowers from a greenhouse or floral shop. Divide the class into small groups, and assign each group to identify and label the parts of their flower using straight pins with labels attached. Rotate the groups around the room until they have inspected the flowers at each station, checking to be sure that the flower parts have been correctly identified. Have each student draw and label the parts of a flower.

3. Obtain several cross-cut sections of a small tree (6–8 inches in diameter). The thickness of each piece should be 2–3 inches to facilitate ease of handling. Assign students to identify the following events in the life of the tree by labeling the appropriate ring with colored or labeled pins:

   A. Year of greatest growth (assuming the tree was cut in the current year)

   B. Drought years or years of greatest stress on the tree

   C. Year of the student's birth

   D. Year of the attack on the World Trade Center

   E. Other significant events as assigned

QUESTIONS FOR DISCUSSION AND REVIEW

Essay Questions

1. What are the basic structures of a plant cell?

   The basic structures of a plant cell include the cell wall, cell membrane, cytoplasm, and nucleus (including DNA).

2. Explain the importance of the process of photosynthesis in sustaining plant and animal life.

   Photosynthesis is the process by which plants convert energy from sunlight to sugar. Sugar is converted to starches and other compounds that the plant stores for later use. Animals obtain their nutrition from stored plant materials by eating the plant or by eating an animal that ate the plant.

3. What are the two high-energy molecules formed during the light reactions as part of the photosynthesis process?

The two high-energy molecules are ATP and NADPH.

4. Name some products that are formed in plants from the simple sugars that are produced during photosynthesis.

Products formed in plants from simple sugars include starches, cellulose, and oils; in combination with nitrogen, proteins may be formed.

5. Describe the formation of starch from simple sugars through the process of dehydration synthesis.

The process of dehydration synthesis occurs as a water molecule is removed from a glucose molecule, causing it to bond to another glucose, starch, or cellulose molecule.

6. Name the type of cell division that accounts for most of the growth in trees, and list the steps in the process.

The type of cell division that accounts for growth in a tree is mitosis. The steps in the process of mitosis include interphase (resting stage), prophase (chromosome formation), metaphase (chromosome alignment), anaphase (chromosome separation), telophase (division of all cell contents), and finally back to interphase.

7. Describe the process of meiosis during the formation of male and female gametes.

The process of meiosis in the formation of male gametes (pollen grains) occurs as follows. Pollen formation begins with the production of microspore mother cells inside the pollen sacs. These are diploid cells that contain chromosome pairs. Each of these cells divides, and a tetrad consisting of a cluster of four haploid cells is formed. These four cells pull apart, forming four microspores. The nucleus of each microspore divides one more time, forming a pollen grain consisting of two cells. One cell contains the generative nucleus, from which two sperm cells will develop during fertilization of the ovules.

Female gametes (ovules) are formed beginning with a megaspore mother cell inside the ovule. It is a diploid cell. This cell divides to form four haploid cells, called megaspores, of which three will die. The remaining megaspore grows in size, and its nucleus divides to form two nuclei. Two more divisions occur, and a total of eight nuclei are produced. These eight haploid nuclei move to different locations in the ovule. Three of the nuclei come together, forming the egg cell. The other nuclei form other plant materials.

8. What are the differences between sexual and asexual propagation of plants?

Sexual propagation of a plant occurs when a pollen grain merges its genetic material with that of the female gamete or egg cell to form a seed. Asexual propagation of plants occurs when a new plant grows from leaf, stem, or root tissue.

9. How is a high forest different from a low forest?

A high forest is different from a low forest because it is propagated from seeds, and a low forest is propagated from roots, stumps, or branches of trees that have been buried beneath the soil surface.

10. Compare the different forms of vegetative reproduction.

The coppice method or sprout method of reproduction regenerates a new forest from sprouts that arise from the roots or stumps of harvested trees. Layering is a form of vegetative reproduction in which live tree branches are buried in the debris on the forest floor, causing roots and stems to be generated.

11. What steps are involved in the propagation of plants using tissue-culture technologies?

Tissue culture is another method of plant propagation in which entire plants are generated from callus tissue by manipulating the plant hormones to cause callus tissue to form roots and stems.

MULTIPLE-CHOICE QUESTIONS

1. A permeable structure found in plant cells that restricts the kind of materials that can enter a cell is the:

    A. <u>cell membrane</u>   C. vacuole

    B. nucleoplasm   D. cell wall

2. Energy from sunlight is captured and stored in plant tissues through the process of:

    A. meiosis   C. dehydration synthesis

    B. <u>photosynthesis</u>   D. mitosis

3. The green substance found in plant cells that plants use to capture energy from sunlight is called:

    A. chloroplast   C. <u>chlorophyll</u>

    B. NADPH   D. cellulose

4. A high-energy molecule formed during the first light reaction is:

    A. <u>ATP</u>   C. cellulose

    B. NADPH   D. chlorophyll

5. The phase of photosynthesis during which carbon dioxide reacts with ATP and NADPH to form simple sugars is:

    A. respiration   C. light reactions

    B. dehydration synthesis   D. <u>Calvin cycle</u>

6. Dehydration synthesis is the plant process responsible for the formation of:

    A. <u>starch</u>   C. cellulose

    B. glucose   D. chloroplast

7. The fats, oils, and waxes in plants are found mostly in:

    A. cellulose   C. <u>seeds</u>

    B. starch   D. lignin

8. Which of the following terms is *not* a stage of mitosis?

    A. prophase   C. telophase

    B. <u>comatose</u>   D. metaphase

9. Meiosis is a form of cell division in which:

    A. <u>gametes are produced</u>   C. plant growth occurs

    B. asexual reproduction occurs   D. callus tissue is formed

10. When sexual reproduction occurs in plants, the name of the male gamete that fertilizes the egg cell is:

    A. anther   C. coppice

    B. <u>pollen</u>   D. chromatid

11. Which of the following terms is *not* associated with propagation of plants from live plant parts?

    A. regeneration   C. vegetative reproduction

    B. asexual reproduction   D. <u>megaspore mother cell</u>

12. In plant tissue culture, a term that describes the plant tissue from which new plants develop is called:

    A. sprouts   C. embryo sac

    B. filament   D. <u>callus</u>

Chapter 17

FOREST ECOLOGY

After completing this chapter, you should be able to

- explain the relationship between soil erosion and pollution of surface water

- understand the basic law of physics known as the *conservation of matter*

- describe how natural cycles function to prevent pollution and renew the environment

- discuss the importance of the element *carbon* to living organisms

- suggest some reasons why soil is considered to be one of our most important natural resources

- distinguish among the three soil orders that are of significance to forestry in North America

- explain the major functions and significance of watersheds

- illustrate the different events that occur in the water cycle

- suggest some effects of air pollution on forests

- describe the similarities and differences among food chains, food webs, and food pyramids

TEACHER BACKGROUND INFORMATION

The science of ecology deals with relationships between organisms and their environments. It is important to understand that living things do not survive alone—that all organisms depend on the environment and other organisms in it for food and favorable living conditions. Most plants depend on soils or water for habitat, and all animals depend on plants for food. Ecology is the science that deals with all of these relationships. Human relationships with the environment are also part of the earth's ecology. This does not mean that we must leave the planet as we found it, but it does mean that we should be careful to study the effects of human activities on the world's environments and the other organisms with which we coexist.

INSTRUCTIONAL STRATEGIES

Gather examples of different soils in gallon containers from a variety of locations. They should be of different textures, colors, structures, and parent material. Make sure that they are distinctly different. These samples should be displayed in the classroom or laboratory. Discuss the differences in the soil samples and encourage the students to feel differences in texture, measure pH, and perform other tests that may distinguish the soil samples from one another. Suggest to the class that all living organisms depend on soil, either directly or indirectly, for survival.

LEARNING ACTIVITIES

1. Following a discussion about food chains, food webs, and food pyramids, arrange for the class to visit an outdoor site near the school. Assign class members to carefully observe and record the plants, insects, and animals that live there. After returning to the classroom, each student should prepare an illustration of a food web that they observed. Display the illustrations in the classroom, and use them in a discussion to determine the order in which the organisms should be arranged to create the food pyramid that most correctly describes the area.

2. Invite a person to visit the class who has expertise in the identification of insects. Extension educators, foresters, and urban foresters in many areas are competent to lead discussions about the ways that insects affect trees and forests. Ask the visitor to use pictures, slides, or actual insect collections to raise the students' interest level. Assign pairs of students to bring a picture or illustration of a particular insect to class. Make a collage on the classroom bulletin board with the pictures and illustrations that the students contribute.

3. Divide the class into groups of two to three students. Allow each group to choose an animal or insect that lives in a forest environment. Each group should prepare a 5-minute oral report on the "ecology" of the creature they selected. They should (1) identify all of the living organisms with which their selected creature interacts, (2) prepare a chart illustrating where the organism fits into the food web, and (3) suggest ways that forests could be managed to improve the survival rate for the animal or insect selected.

QUESTIONS FOR DISCUSSION AND REVIEW

Essay Questions

1. How does soil erosion contribute to the pollution of surface water?

   Soil particles suspended in surface water are the primary cause of water pollution. Water that moves rapidly across the soil surface disturbs silt particles, which become suspended in the water until it slows down enough to allow them to settle to the bottom.

2. Explain the basic principle of physics known as the law of the conservation of matter.

   The law of the conservation of matter states that matter may change from one form to another, but it cannot be created or destroyed by natural physical or chemical processes. Nothing is ever lost in nature. We can change the form of waste materials, but we cannot eliminate them.

3. What are the key elements of the carbon and nitrogen cycles, and how do such cycles prevent pollution of the environment?

The carbon and nitrogen cycles prevent pollution by changing the form of carbon and nitrogen compounds. The carbon cycle occurs as carbon moves readily between living organisms, the atmosphere, the ocean, and the soil. The nitrogen cycle occurs as nitrogen gas in the atmosphere is formed into compounds in the soil that can be absorbed as nutrients for plants. Nitrogen is part of the protein molecules that are important nutrients for animals. It is returned to the atmosphere and the soil when plants or animals die and decompose.

4. What makes carbon so important to living organisms?

Carbon is important to living organisms because it is the most abundant element in the dry matter of plants and animals. It makes up the framework of the molecules found in living tissue.

5. Why is soil considered to be one of our most important natural resources?

Soil is considered to be one of our most important natural resources because it is the source of the nutrients required by all living organisms to live, grow, and reproduce.

6. What are the three most important classes of forest soils, and what characteristics separate them into distinct classes?

The three most important classes of forest soils and their characteristics are:

- Alfisols: clay soils high in calcium, magnesium, sodium, and potassium, which tend to be slightly acid
- Spodosols: soils found in cold, damp climates; they are formed of coarse silica parent material, light in color, acid pH, illuvial (humus, aluminum, and iron have leached into the B, A, and E horizons)
- Ultisols: soils found in warm, humid climates; they show evidence of heavy weather action with illuvial deposits of red and yellow clay in the B horizon

7. How does a watershed function, and why are healthy watersheds important?

A healthy watershed is important because it absorbs excess water from heavy precipitation or during periods of heavy snowmelt. This water percolates slowly down through the sand, gravel, and rocks until it surfaces through free-flowing springs of fresh water.

8. Describe each of the events known to occur as the water cycle functions.

The events that occur as the water cycle functions include the following steps. Water molecules evaporate from the ocean and other bodies of water and fall as precipitation on land masses. Some of the water is absorbed by a watershed, but all water flows toward the ocean through streams and rivers in response to the force of gravity. Some water is taken up by plants or animals, but eventually it passes back to the ocean to begin the cycle all over again.

9. In what ways does polluted air affect forests?

Polluted air containing sulfur dioxide or nitric oxide forms weak acids when these pollutants come in contact with precipitation. These acids are capable of causing severe damage and even death to trees. Acid precipitation also pollutes forest water sources on which all of the organisms in the forest environment depend.

10. How are food chains, food webs, and food pyramids similar? What are the differences among them?

A food chain begins with a producer or food plant. Food plants are consumed by animals known as primary consumers. These animals are eaten by predatory animals (secondary consumers), with the most dominant predator at the top of the food chain. A food web consists of two or more overlapping and interwoven food chains. A food pyramid arranges creatures in a ranking order according to each organism's dominance in a food web.

MULTIPLE-CHOICE QUESTIONS

1. The source of all energy used by plants and animals is:

   A. water   C. soil

   B. <u>sunlight</u>   D. air

2. The greatest single pollutant of surface water is:

   A. <u>silt</u>   C. industrial waste

   B. air pollution   D. mine tailings

3. Acid precipitation occurs when raindrops are converted to weak acids by absorbing:

   A. industrial waste   C. <u>sulfur dioxide</u>

   B. animal waste   D. silt

4. Which natural cycle is *not* an elemental cycle?

   A. carbon cycle   C. <u>water cycle</u>

   B. nitrogen cycle   D. oxygen cycle

5. A process by which bacteria are able to convert atmospheric nitrogen into nitrates useful to plants is called:

   A. <u>nitrogen fixation</u>   C. denitrification

   B. nitrogenization   D. metabolism

6. A process by which a buildup of translocated soil components occurs is known as:

   A. eluviation   C. erosion

   B. particle concentration   D. <u>illuviation</u>

7. A soil order that is high in concentrations of calcium, magnesium, sodium, and potassium is:

   A. <u>Alfisol</u>   C. Spodosol

   B. Ultisol   D. humus

8. A destructive process that sometimes occurs in soils that are not protected against forces of flowing water or strong winds is:

   A. denitrification   C. <u>erosion</u>

   B. conservation of matter   D. soil conservation

9. A common description of a watershed is:

   A. an indoor toilet

   B. a small building that protects a well equipped with a water pump

   C. <u>an area in which rain and melting snow are absorbed to emerge as springs at lower elevations</u>

   D. a swampy lowland area

10. Rain or snow that is contaminated with smoke from engine exhausts or industrial plants forms a destructive product called:

    A. <u>acid precipitation</u>   C. smog

    B. water cycle   D. transpiration

11. What is another name for a predatory animal that eats other animals?

    A. <u>secondary consumer</u>   C. herbivore

    B. producer   D. primary consumer

12. What or who usually occupies the dominant position in a food pyramid?

    A. an omnivore   C. a primary consumer

    B. <u>a predator</u>   D. a herbivore

Chapter 18

DISEASES AND PESTS OF TREES

After completing this chapter, you should be able to

- distinguish between biotic and abiotic diseases
- describe the symptoms of the heart-rot diseases, and explain the significance of these diseases for the forest products industry
- explain the roles that fungi play in causing diseases of trees
- describe the symptoms of the canker diseases, and identify some control methods that are used in the forest industry
- describe the symptoms of the rust diseases, and identify some control methods that are used in the forest industry
- identify the symptoms of vascular wilt infections in trees, and recommend a method for controlling these diseases

- name and describe some forms of abiotic diseases in trees
- name the most important classes of destructive forest insects, and describe the nature of the damage that each inflicts upon trees
- distinguish between biological control and chemical control of insects, and list some examples of each control method
- discuss the merits of integrated pest management (IPM) and genetic engineering as insect control methods for the future
- identify other pests that damage or kill trees, and suggest methods for controlling them

TEACHER BACKGROUND INFORMATION

At least two principles are involved in decisions concerning the control of diseases and pests in forests. The first has to do with whether the value of the trees saved is sufficient to offset the cost of the treatment. This is an economic decision based on the law of diminishing returns. The second principle comes into play each time forest managers contemplate a widespread chemical treatment. Some people believe that it is unethical to resort to any chemical treatments for any purpose on forests and other public lands. Others believe that it is not morally right to allow a pest infestation to spread when a method of control is available. Each of these viewpoints is probably expressed in most classrooms. Teachers should help students obtain the facts that will let them make up their own minds on such controversial issues without imposing the teacher's viewpoint.

INSTRUCTIONAL STRATEGIES

A fun-filled approach to teaching this lesson would be to conduct a mock trial of one of the students on charges of polluting the environment and killing honeybees with a chemical application intended to control beetles in an adjacent pine forest. Two or three students might be selected as a prosecution team, with another team to defend the student who is bound over for trial. A jury should be selected, and the rest of the class might be called as witnesses. Each student should have a set of instructions that clearly defines his or her role. The teacher acts as the judge in this setting and controls the course of events.

Students usually display a great deal of imagination when this teaching strategy is used, and they will be motivated to study controversial issues. Students will benefit from the arguments of their classmates on key issues while enjoying an interesting learning experience on what might otherwise be a dull subject.

LEARNING ACTIVITIES

1. Take a walking field trip around the neighborhood and collect insect specimens suspected of causing damage in trees. Identify each insect that has been collected, and study how each insect interacts with the trees. Suggest ways that each insect might be controlled to prevent damage to trees. Assign students to repeat this exercise by gathering and displaying their own insect specimen.

2. Invite the county extension educator or a forest or park service official to discuss local forest and ornamental tree problems with members of the class. Discuss procedures for working with plant materials from diseased plants that will prevent the diseases from spreading to new areas.

3. Identify several insects and diseases that affect forests in your area. Assign students in groups of two or three to research one of these insect or disease problems to identify the following: (a) What kind of damage does it cause? (b) What methods of control have been used effectively? (c) What are some new or emerging solutions that may be used to control the insect or disease? (Students may report their findings in written and/or oral reports.)

QUESTIONS FOR DISCUSSION AND REVIEW

Essay Questions

1. How are biotic diseases of plants different from abiotic diseases?

   Biotic diseases of plants are caused by living agents of infection, such as bacteria, fungi, viruses, micoplasmas, parasites, and nematodes. An abiotic disease is caused by a nonliving factor or condition, such as a chemical injury.

2. Describe the symptoms of the heart-rot diseases, and explain their significance to the forest products industry.

   Symptoms of the heart-rot diseases include decay of the core of deadwood in the center of a mature tree and loss of the flow of sap. Affected trees become hollow and weak, and they are vulnerable to wind. Heart-rot diseases cause greater economic losses in the North American forest industry than any other type of disease.

3. What roles do fungi play in causing diseases of trees?

The most important and destructive disease agents in the forests are the fungi. They contribute to root rot, heart-rot, cankers, and stains. All of these disease symptoms can be traced back to fungus infections.

4. What are the disease symptoms of canker diseases in trees? How are these diseases controlled?

The symptoms of canker diseases include infections of the bark and cambium layers. These girdle trees, stop the flow of nutrients, and cause the death of infected trees. Perennial cankers cause lesions on the edges of infected areas. Growths that cause damage to the woody tissues also occur. Cankers are best controlled by removing infected trees from the forest and burning them.

5. Describe the symptoms of the rust diseases, and identify some control methods used in the forest industry.

Symptoms of the rust diseases include spotted red or brown discoloration of the stems and leaves of trees. Rust infections affect trees differently from one variety to another. For example, the cones, needles, and stems of some conifer trees may be affected, whereas only the leaves of certain hardwoods become diseased. All of the fungi that cause rust diseases spend part of their life cycles on unrelated host plants. The best control methods for rust diseases include the use of fungicides in forest nurseries and planting trees that are resistant to the fungus infection.

6. Identify the symptoms of vascular wilt infections in trees, and recommend a method for controlling these diseases.

The symptoms of vascular wilt infections include wilting of the leaves, followed by death of some of the branches. The disease is spread by insects and through root grafts that connect adjacent trees. These diseases can sometimes be prevented by trenching to eliminate root grafts, avoiding injuries to trees, injecting threatened trees with fungicides, and burning infected trees. The best solution is to plant disease-resistant trees.

7. Name and describe some forms of abiotic diseases in trees.

Forms of abiotic diseases in trees include sunscald, blight (death of foliage), poisoning (chemicals or pollutants), nutrient deficiencies, mechanical damage, and weather-related damage.

8. What classes of insects are harmful to forests, and what kinds of damage does each insect class inflict on trees?

The classes of insects that inflict damage on forests and the kinds of damage include the following:

- Bark beetles: create tunnels between the bark and the woody part of the tree trunk that eventually girdle the tree
- Defoliators: feed on the leaves and needles of trees
- Root-feeding insects: eat and destroy root tissue
- Terminal-feeding insects: eat meristem tissue in the tips of branches and the central leaders of infected trees
- Sucking insects: feed on the resin and sap of trees
- Wood borers: bore holes through the sapwood and heartwood of mature trees

9. Describe and give examples of biological and chemical control methods for forest insects.

Biological control methods for insects include the introduction of insect predators into the forest environment. These predators may eat insect pests as adults, or they may lay their eggs near or inside the bodies of harmful insects so the larva that hatches will have a food supply. Biological control also includes the introduction of disease-causing organisms into the environment that specifically affect the insect pests. Chemical control of insect pests is achieved by treating an insect-infested area with recommended amounts of insecticides. The more specific the insecticide is to the pest, the more useful it is in the forest.

10. How are genetic engineering and integrated pest management practices used to control insects?

New tools for insect control include the use of genetic engineering practices to develop insect-resistant trees. This practice, when used in conjunction with biological, chemical, and mechanical forms of insect control, constitutes a new approach that is called integrated pest management. This approach provides the best protection to useful insects while targeting harmful insects.

11. What other kinds of pests damage or kill trees, and how can they be controlled?

Other pests that injure or kill trees include rodents such as mice, gophers, and porcupines, to name a few. Control of these pests is accomplished using poison baits. Deer are sometimes destructive to young trees. Deer can sometimes be repelled by placing muslin bags of blood meal in the branches of vulnerable trees and bushes. Depredation hunts are used to harvest game animals that persist in causing damage to forests.

MULTIPLE-CHOICE QUESTIONS

1. What is a disease that is caused by a living organism called?

   A. not contagious   C. pollution

   B. abiotic   D. <u>biotic</u>

2. Wood rots are diseases that occur in trees as a result of organisms called:

   A. <u>fungi</u>   C. rhizomorphs

   B. cankers   D. rusts

3. Fungi are organisms that lack chlorophyll to make their own food, so they obtain nourishment from other living things. For this reason, fungi are known as:

   A. endomorphs   C. <u>parasites</u>

   B. rhizomorphs   D. cankers

4. When the roots of trees touch each other, they sometimes grow together, creating connections between their roots that are
known as:

   A. rusts   C. wilts

   B. <u>root grafts</u>   D. rhizomorphs

5. Which disease is known as a wilt?

   A. white pine blister rust   C. heart rot

   B. <u>Dutch Elm disease</u>   D. dwarf mistletoe

6. A fruiting body is:

   A. <u>a structure that produces spores that develop into fungi</u>

   B. a fleshy structure that surrounds the seeds of a plant

   C. a thin strand of fungal tissue that enters tree roots, infecting them with disease organisms

   D. a structure on a tree leaf in which sap becomes fermented to produce honeydew

7. A canker is a plant infection that:

   A. causes red or brown discoloration

   B. affects only conifer trees

   C. <u>kills patches of tissue on the trunk or branches</u>

   D. is a painful infection of the gum of a tree

8. An example of an abiotic disease in a tree is:

   A. canker   C. mistletoe

   B. verticillium wilt   D. <u>sunscald</u>

9. An entomologist is:

   A. a person who studies relationships between living and
       nonliving things

   B. a student of the fruiting habits of trees

   C. a hollow, beaklike mouthpart with which sucking insects obtain sap for food

   D. <u>a person who studies the branch of science related to insects</u>

10. Which of the following is *not* a destructive type of forest insect?

    A. defoliator   C. bark beetle

    B. terminal feeder   D. <u>pollinator</u>

11. To which of the following types of destructive forest insects does the scale insect belong?

    A. <u>sucking insect</u>   C. wood borer

    B. defoliator   D. root feeder

12. Which of the following animals is *not* classed as a rodent?

    A. gopher   C. mouse

    B. <u>mink</u>   D. porcupine

Chapter 19

ANATOMY AND CLASSIFICATION OF TREES

After completing this chapter, you should be able to

- define *dendrology*
- distinguish between the anatomy and the physiology of a tree
- describe the different tissue systems of a tree
- explain the importance of xylem and phloem tissues in a tree
- describe the importance of meristem tissue as it relates to the growth of trees
- identify the external parts of a tree leaf
- illustrate the basic structure of a tree root
- name the basic parts of a flower

TEACHER BACKGROUND INFORMATION

The information contained in this chapter is covered in detail in most good biology or botany textbooks. It would be a good idea to use such a book as a reference as you teach this chapter. Most students should be familiar with some of these terms from science classes they have taken in the past, but very few will have experience in identifying trees from their leaf structure.

INSTRUCTIONAL STRATEGIES

Cut enough cross-section pieces from a fresh tree limb or small tree for each pair of students to have a specimen. Point out the cambium layer, xylem, phloem, and bark and discuss how a tree grows in thickness. Use this teaching tool to introduce this chapter on anatomy.

LEARNING ACTIVITIES

1. Examine the scale-leaf structure of a juniper, cedar, or cypress tree (ornamental shrub). Assign the students to draw what they see. Speculate with the class about how this leaf structure is able to conserve water. Name some other life forms found in desert environments that have scales, and consider whether they function in a similar way.

2. Obtain the flowers of several different types of woody plants and examine them under magnification. Sketch the flowers, and identify the male and female structures of each. Assign a class member to describe the role of each of these structures.

QUESTIONS FOR DISCUSSION AND REVIEW

Essay Questions

1. What is the difference between the anatomy and the physiology of a tree?

   The anatomy of a tree is a study of the arrangement and relationships of the various organs to each other. The physiology of a tree is a study of the life functions and life processes that occur in a tree.

2. Describe the different tissue systems of a tree.

   The tissue systems of a tree include the following:
   • Ground tissue: parenchyma, collenchyma, and sclerenchyma cells
   • Vascular tissue: water-conducting tissue such as xylem, tracheids, and sieve tubes
   • Dermal tissue: protective tissues such as epidermis, cuticle, and cork
   • Meristem: cambium and apical meristem tissue

3. How is xylem tissue adapted to its function of transporting dissolved materials within a tree?

   Xylem tissue is adapted to transporting dissolved materials because it is made of cells that are hollow or that form passages through which water and dissolved nutrients can flow.

4. What role does meristem tissue play in the growth of a tree?

   The role of meristem tissue in the growth of a tree is to deposit wood to the outer layer of the xylem and to the inner layer of the bark from the cambium tissue. Vertical growth occurs from the apical meristem of the tree, resulting in the tree becoming taller.

5. What are the external structures of a leaf, and what are their purposes?

   The external structures of a leaf and their purposes are:
   • Petiole: point of attachment between the leaf and the tree; consists of vascular tissues that transport water, minerals, and nutrients to the leaf cells
   • Blade: the flat part of the leaf that collects sunlight
   • Midrib: gives strength to the leaf and includes vascular tissue for the transfer of nutrients

- Spines: same function as the midrib
- Veins: distribute materials between spines or midrib and cells of a leaf
- Margin: useful in identifying the species of a tree

6. What are the basic structures of a tree root?

The basic structures of a tree root include the vascular cylinder, root hairs, procambium, cortex, epidermis, endodermis, pericycle, cork cambium, and root cap.

7. How are compound leaves different from simple leaves?

A simple leaf is one that has only one set of leaf parts. It has a single blade and only one petiole. A compound leaf consists of three or more leaflets.

8. What major difference exists between trees classed as angiosperms and those classified as gymnosperms?

Trees that bear seeds in cones are called gymnosperms. Trees that produce seeds in ovaries and fruits are called angiosperms.

9. Name and describe the basic functions of the parts of male and female flowers.

The basic parts and functions of male and female flowers include:

- Female: Stigma—functions as a pollen receptor

   Ovary—produces the egg cell or ovum
   Style—connects the stigma to the ovary

- Male: Anther—produces pollen grains

   Receptacle—base of the flower
   Filament—connects the anther to the receptacle

10. What are the names and distinguishing characteristics of six groups of trees classified by the National Audubon Society according to the structure of their leaves?

The names and distinguishing characteristics of the tree groups classified by the National Audubon Society according to leaf structure include the following:

- Needle-leaf conifer: leaves are long in length and narrow in width
- Scale-leaf conifer: leaves are shaped like tiny overlapping scales
- Unserrated simple leaf: single blade and only one petiole with a smooth leaf margin
- Toothed simple leaf: single blade and only one petiole with a sawtooth-shaped leaf margin
- Lobed simple leaf: single blade and only one petiole with the appearance of rounded divisions along the leaf margins
- Compound leaf: multiple leaflets attached at the end of a single leafstalk

MULTIPLE-CHOICE QUESTIONS

1. What is a study of the structure of an organism called?

   A. physiology   C. anatomy

   B. permeability   D. taxonomy

2. Which system is *not* one of the basic tissue systems of a plant?

   A. vascular tissue   C. ground tissue

   B. nucleoplasm   D. dermal tissue

3. A plant cell that has thick cell walls which add strength to plant stalks and stems is called:

   A. parenchyma   C. collenchyma

   B. nucleoplasm   D. sclerenchyma

4. Phloem is a conductive tissue that includes which type of structure?

    A. <u>sieve tube</u>   C. vessel element

    B. tracheid   D. sclerenchyma

5. A plant structure that transports dissolved materials across the woody section of a stem is called:

    A. apical meristem   C. sieve tube

    B. pith   D. <u>vascular ray</u>

6. The petiole is a plant structure found in a:

    A. flower   C. stem

    B. <u>leaf</u>   D. root

7. A plant root tissue that stores starches is:

    A. <u>cortex</u>   C. endodermis

    B. cambium   D. epidermis

8. A female flower part in which the seed forms is the:

    A. <u>ovule</u>   C. receptacle

    B. stamen   D. style

9. What is the male flower part in which pollen grains develop and mature?

    A. filament   C. stigma

    B. <u>anther</u>   D. sepal

Section Seven

TREES OF THE FOREST

Chapter 20
TREE IDENTIFICATION

OBJECTIVES

After completing this chapter, you should be able to

- recognize characteristics of leaves, bark, flowers, and seeds
- observe and recognize the growth patterns of trees
- associate tree growth patterns with specific trees
- identify each tree that is featured in this chapter

TEACHER BACKGROUND INFORMATION

You should carefully review Chapter 19 of the textbook in preparation for teaching this chapter. Pay particular attention to the different leaf structures illustrated there. Obtain a good field guide for trees in your region of the country and learn to detect the differences between trees of similar anatomy and form. A good identification key for trees would be most helpful as students bring leaves, bark, flowers, and seeds to class, seeking your guidance to properly identify them.

INSTRUCTIONAL STRATEGIES

Assemble a collection of tree parts, including leaves, flowers, bark, and seeds. These samples should be used for the purpose of teaching students to identify the common and scientific names of trees. Provide incentives for the students to learn to identify the names of the trees from which the samples were obtained. Assign students to assemble and submit their own collections of plant samples for identification purposes. Test daily during the time that students are learning to identify trees.

LEARNING ACTIVITIES

1. Study and memorize tree identification characteristics, such as seed, leaf, and bark patterns. Many of these unique characteristics are illustrated with photographs for each of the trees included in this chapter. Repetition and careful observation are the keys to identifying trees by name.

2. Provide time for study of Chapter 20 each day for several days, tracking each student's progress towards mastery.

3. Use images and samples of tree parts to test students' abilities to correctly identify common and scientific names of the tree species found in your area.

NOTES

NOTES

NOTES

NOTES

NOTES

NOTES